10 North Dakota NDSA Grade 5 Math Practice Tests

The Ultimate Test Prep Collection with Answer Explanations

Dr. A. Nazari

10 Practice Tests

The Complete Mastery Collection

Welcome, young scholar!

*You hold in your hands the **complete collection** —
ten full practice tests to make you a true*
Math Master.

- ✳ *Master every question type*
- ✳ *Build unshakeable confidence*
- ✳ *Track your rise from beginner to expert*
- ✳ *Arrive at test day fully prepared*

Your journey to mastery begins now.

> **❝** *Ten tests may sound like a lot, but each one makes the next one easier. Trust the process!* **❞**

The Master Plan

Your 4-phase journey from apprentice to master

Phase I: Discovery (Tests 1–3)

Explore the test format. No timing, no stress. Read the answer explanations after each test to understand every question.

Phase II: Growth (Tests 4–6)

Set a gentle timer (75 minutes). Focus on the topics you found hardest. Practice showing your work on every problem.

Phase III: Strength (Tests 7–9)

Full timed conditions (60 minutes). Simulate the real exam. Review only the questions you missed — don't re-read what you already know.

Phase IV: Mastery (Test 10)

Your final challenge. Full exam conditions. This is your victory lap — show yourself how far you've come!

Question Types in Each Test

- **Multiple Choice** — select the best answer
- **Short Answer** — show your work
- **Multi-Select** — choose ALL correct answers
- **Open Response** — explain your reasoning

Scholar's Note: Space your tests 2–3 days apart. Use the days between for review. By Test 10, you'll be amazed at your growth.

 The Master's Wisdom

Seven rules that separate beginners from masters

I   **Read the full question twice.** The first read tells you what it's about. The second read tells you what to do.

II   **Mark the important parts.** Circle key numbers, underline what the question asks, and cross out information you don't need.

III   **Decide your operation.** Before solving, ask: Am I adding? Subtracting? Multiplying? Dividing? Write it down.

IV   **Solve on scratch paper first.** Even for multiple choice — work it out, then look at the answer choices.

V   **Eliminate wrong answers.** On multiple choice, cross out answers that are clearly wrong. This doubles your chances.

VI   **Check with estimation.** After solving, estimate the answer mentally. If your exact answer is far from the estimate, recheck.

VII   **Never leave a question blank.** Even your best guess is better than no answer. Use what you know to make an educated choice.

🕐 Timing Mastery

Tests 1–3: **Untimed** (learn the format) › Tests 4–6: **75 min** (build speed) › Tests 7–10: **60 min** (exam conditions)

 66 *A true master isn't someone who never makes mistakes — it's someone who learns from **every single one**. After each test, review your errors. That's where the real learning happens.* 99

The Scholar's Toolkit

Prepare your workspace before each session

✏️	**Sharpened Pencils**	Two #2 pencils — always have a backup ready
	Quality Eraser	A clean, soft eraser that won't smudge your work
	Scratch Paper	Blank paper for calculations and diagrams
	Ruler	Helpful for measurement and geometry questions
	Timer	Begin using from Phase II onward
	Quiet Workspace	A calm, well-lit area free from distractions

Permitted on Exam Day

- ✔ Pencil and eraser
- ✔ Scratch paper (provided)
- ✔ Ruler (if specified)

Not Permitted

- ✗ Calculator
- ✗ Electronic devices
- ✗ Reference materials

♥ A Note for Parents & Guardians

Ten practice tests provide the most thorough preparation available. The 4-phase structure (Discovery → Growth → Strength → Mastery) ensures your child builds skills progressively.

Keys to success:

- Space tests 2–3 days apart — never more than one per day
- Phases I–II: sit with your child and review answers together
- Phases III–IV: let them work independently, then discuss results
- Celebrate the journey — compare Test 1 to Test 10

Find more at
ViewMath.com/ND-Grade5

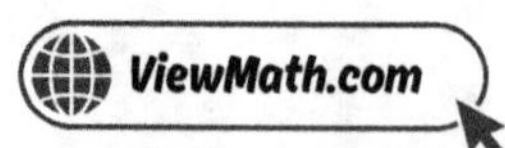

Math Reference Sheet

You may use this page during your practice tests!

Symbol	Name	What It Means	
$(\ )$	Parentheses	Do this part first.	$(3+4) \times 2 = 14$
10^3	Exponent	Multiply the base by itself that many times. $10^3 = 1,000$	
$\dfrac{a}{b}$	Fraction	a parts out of b equal parts; also means $a \div b$.	
$\dfrac{7}{3}$	Improper Fraction	Numerator $\geq$ denominator.	$\dfrac{7}{3} = 2\tfrac{1}{3}$
0.45	Decimal	A number with a decimal point.	$0.45 = \dfrac{45}{100}$
$> < =$	Comparison	Greater than, less than, equal to.	$0.5 > 0.35$
$(3,5)$	Ordered Pair	A point on the coordinate plane: (x, y).	

🖩 Key Formulas

- **Volume of a rectangular prism:**

  $V = l \times w \times h$

- **Order of operations:**

  Parentheses $\rightarrow$ Exponents $\rightarrow$ Multiply/Divide $\rightarrow$ Add/Subtract

- **Powers of 10:**

  $10^1 = 10 \quad 10^2 = 100$

  $10^3 = 1,000 \quad 10^4 = 10,000$

- **Fraction as division:**

  $\dfrac{a}{b} = a \div b$

⊞ Place Value Chart

Millions	1,000,000	**Decimals**	
Hundred-Thousands	100,000	Tenths	0.1
Ten-Thousands	10,000	Hundredths	0.01
Thousands	1,000	Thousandths	0.001
Hundreds	100		
Tens	10		
Ones	1		

Each place is **10**× the place to its right, and $\frac{1}{10}$ of the place to its left.

🗒 Key Math Vocabulary

- **Sum** — the result of addition
- **Difference** — the result of subtraction
- **Product** — the result of multiplication
- **Quotient** — the result of division
- **Remainder** — what's left over after dividing
- **Factor** — a number you multiply
- **Expression** — numbers and operations without =
- **Equation** — a math sentence with =
- **Numerator** — the top number of a fraction
- **Denominator** — the bottom number of a fraction
- **Mixed number** — a whole number + a fraction
- **Equivalent fractions** — fractions with equal value
- **Decimal** — a number written with a decimal point
- **Volume** — the space inside a 3D shape
- **Coordinate plane** — a grid with x and y axes
- **Ordered pair** — (x, y) location on the plane

🔍 Word Problem Clue Words

- **Add** (+): total, altogether, combined, sum, increase, more than
- **Subtract** (−): difference, how many more, fewer, remain, decrease, left
- **Multiply** (×): each, every, groups of, times, product, per, of (with fractions)
- **Divide** (÷): share equally, split, each group, how many groups, quotient, per

Find more at
ViewMath.com/ND-Grade5

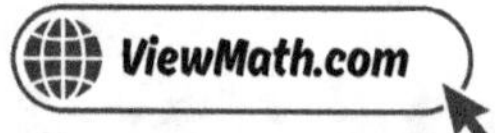

⊞ Multiplication Table ⊞

×	1	2	3	4	5	6	7	8	9	10	11
1	1	2	3	4	5	6	7	8	9	10	11
2	2	4	6	8	10	12	14	16	18	20	22
3	3	6	9	12	15	18	21	24	27	30	33
4	4	8	12	16	20	24	28	32	36	40	44
5	5	10	15	20	25	30	35	40	45	50	55
6	6	12	18	24	30	36	42	48	54	60	66
7	7	14	21	28	35	42	49	56	63	70	77
8	8	16	24	32	40	48	56	64	72	80	88
9	9	18	27	36	45	54	63	72	81	90	99
10	10	20	30	40	50	60	70	80	90	100	110
11	11	22	33	44	55	66	77	88	99	110	121

💡 How to Use This Table

To find **4 × 7**:

1. Find **4** in the left column (blue).
2. Find **7** in the top row (blue).
3. Follow the row and column until they meet: the answer is **28**!

ⓘ **Tip:** You can also use this table for division! If you know 28 ÷ 4 = ?, find 28 in the 4's row. The column header gives you the answer: **7**!

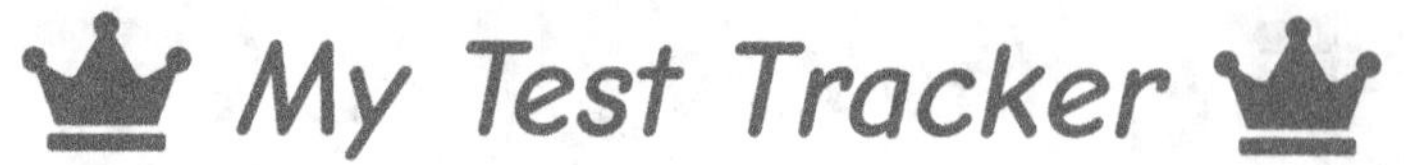 My Test Tracker

Record your scores and watch yourself level up!

Name: _______________________________   **Start Date:** _______________

⭐ Test 1	Date: __________	Score: _____ / 30	%: _______	⭐⭐⭐⭐⭐
⭐ Test 2	Date: __________	Score: _____ / 30	%: _______	⭐⭐⭐⭐⭐
⭐ Test 3	Date: __________	Score: _____ / 30	%: _______	⭐⭐⭐⭐⭐

⭐ Test 4	Date: __________	Score: _____ / 30	%: _______	⭐⭐⭐⭐⭐
⭐ Test 5	Date: __________	Score: _____ / 30	%: _______	⭐⭐⭐⭐⭐
⭐ Test 6	Date: __________	Score: _____ / 30	%: _______	⭐⭐⭐⭐⭐

⭐ Test 7	Date: __________	Score: _____ / 30	%: _______	⭐⭐⭐⭐⭐
⭐ Test 8	Date: __________	Score: _____ / 30	%: _______	⭐⭐⭐⭐⭐
⭐ Test 9	Date: __________	Score: _____ / 30	%: _______	⭐⭐⭐⭐⭐

| ⭐ Test 10 | Date: __________ | Score: _____ / 30 | %: _______ | ⭐⭐⭐⭐⭐ |

Your Rank: ✹ Apprentice (0–39%) ✹ Scholar (40–59%) ✹ Expert (60–79%) 👑 *Master* (80–100%)

Color the stars after each test to track your rank!

📜 Topics to Study More

✹ _______________________

✹ _______________________

✹ _______________________

✹ _______________________

👑 Topics I've Mastered

★ _______________________

★ _______________________

★ _______________________

★ _______________________

Find more at
ViewMath.com/ND-Grade5

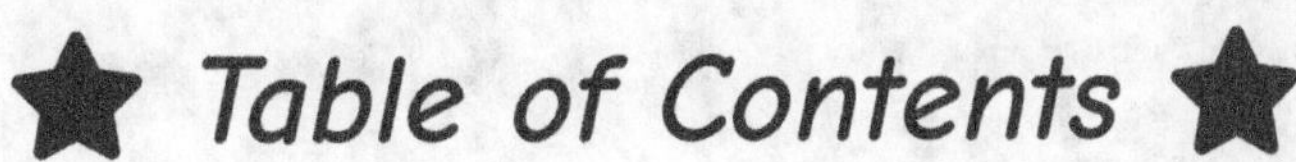 ★ Table of Contents ★

Here's what we'll explore together!

 *Let's learn and have fun!*

Practice Test 1

 30 Questions

✏ Before You Start ✏

- ✔ **Read each question carefully** before choosing your answer.
- ✔ **Show your work** on scratch paper when you need to.
- ✔ **Skip hard questions** and come back to them later.
- ✔ **Check your answers** when you're done.
- ✔ **Take your time** — there's no rush!

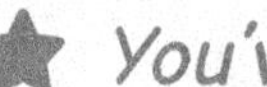 You've Got This! 

Do your best and show what you know!

1. *What is the value of 10^3?*

   (A) 30   (B) 300

   (C) 1,000   (D) 10,000

2. *$0.47 = \dfrac{?}{100}$. What is the missing numerator?*

   (A) $\dfrac{47}{10}$ (B) $\dfrac{47}{1,000}$

   (C) $\dfrac{47}{100}$ (D) $\dfrac{4}{7}$

3. *Round 7.349 to the nearest tenth.*

   Your Answer:

4. *Which expression shows the area model for 52×47?*

   (A) $(50 \times 40) + (2 \times 7)$ (B) $(50 \times 40) + (50 \times 7) + (2 \times 40) + (2 \times 7)$

   (C) $(50 \times 7) + (2 \times 40)$ (D) $(52 \times 40) + (52 \times 70)$

5. *A library has 425 books to pack into boxes. Each box holds 30 books. How many full boxes will there be?*

   (A) 14   (B) 15

   (C) 5   (D) 13

6. *Estimate the quotient of $4{,}780 \div 58$ using compatible numbers.*

   Your Answer:

Find more at
ViewMath.com/ND-Grade5

ViewMath.com

7. *Subtract:* $8.3 - 5.17$

(A) 3.23   (B) 3.87

(C) 3.13   (D) 2.13

8. *Pencils cost $0.75 each. How much do 8 pencils cost?*

(A) \$5.00   (B) \$5.60

(C) \$6.00   (D) \$7.50

9. *What is $6.3 \div 0.9$?*

Your Answer:

10. *Add:* $\frac{4}{9} + \frac{1}{3}$. *Write your answer in simplest form.*

Your Answer:

11. *A student says $\frac{2}{5} + \frac{1}{2} = \frac{3}{7}$. Use estimation to check. Is the answer reasonable?*

(A) *Yes, because $\frac{3}{7}$ is close to $\frac{1}{2}$.*   (B) *No, because the estimate is about 1 but $\frac{3}{7}$ is less than $\frac{1}{2}$.*

(C) *Yes, because $3 + 7 = 10$.*   (D) *No, because $\frac{3}{7}$ is greater than 1.*

12. *Which clue word tells you to subtract?*

(A) total   (B) combined

(C) altogether   (D) remaining

Find more at
ViewMath.com/ND-Grade5

ViewMath.com

13. Which division problem does $\frac{5}{8}$ represent?

(A) $8 \div 5$

(B) $5 \div 8$

(C) 5×8

(D) $8 - 5$

14. Which multiplication gives a whole number answer?

(A) $\frac{1}{3} \times 5$

(B) $\frac{2}{5} \times 7$

(C) $\frac{3}{7} \times 4$

(D) $\frac{1}{4} \times 12$

15. A rectangle is $\frac{3}{4}$ inch long and $\frac{2}{5}$ inch wide. What is its area?

(A) $\frac{5}{9}$ square inch

(B) $\frac{6}{20}$ square inch

(C) $\frac{3}{10}$ square inch

(D) $\frac{6}{9}$ square inch

16. Multiplying a number by $\frac{1}{2}$ gives the same result as:

(A) Adding $\frac{1}{2}$ to the number.

(B) Dividing the number by 2.

(C) Subtracting $\frac{1}{2}$ from the number.

(D) Multiplying the number by 2.

17. What is $2\frac{1}{2} \times 1\frac{1}{5}$?

Your Answer

18. A store originally has 60 apples. On Monday, $\frac{1}{3}$ are sold. How many apples remain after Monday?

(A) 20

(B) 30

(C) 40

(D) 45

Find more at
ViewMath.com/ND-Grade5

19. What is $\frac{1}{4} \div 5$?

(A) $\frac{1}{9}$

(B) $\frac{5}{4}$

(C) $\frac{1}{20}$

(D) $\frac{4}{5}$

20. Evaluate: $[20 - (4 \times 3)] + 7$

(A) 23

(B) 11

(C) 15

(D) 19

21. Write the expression $(24 \div 8) + 5$ as a word phrase.

Your Answer:

22. Rule A: Start at 0, add 3. Rule B: Start at 0, add 12. Write the first 4 ordered pairs (A, B).

Your Answer:

23. Which point is farthest to the right on the coordinate plane?

(A) $(3, 7)$

(B) $(9, 1)$

(C) $(5, 5)$

(D) $(7, 8)$

24. Plot $(1, 1)$, $(1, 4)$, $(6, 4)$, $(6, 1)$ and connect them in order. What shape do they form?

(A) Triangle

(B) Square

(C) Rectangle

(D) Pentagon

Find more at
ViewMath.com/ND-Grade5

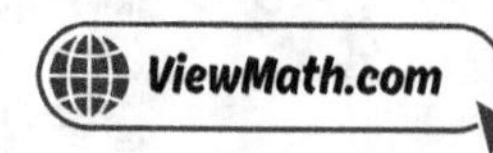

25. Convert 15 feet to yards.

Your Answer:

26. Which question can you answer using a line plot?

(A) What color is the most popular?   (B) How many data points have the same value?

(C) What is the area of the graph?   (D) What is the perimeter of the number line?

27. A rectangular prism has 6 cubes in each row, 2 rows in each layer, and 4 layers. What is its volume?

(A) 12 cubic units   (B) 24 cubic units

(C) 36 cubic units   (D) 48 cubic units

28. A rectangular block is $14 \times 10 \times 5$ cm. A $4 \times 4 \times 5$ cm corner is removed. What is the remaining volume?

Your Answer:

29. A rectangular pool is 15 m long, 6 m wide, and 2 m deep. How many cubic meters of water does it hold?

Your Answer:

30. I have 4 sides. Both pairs of opposite sides are parallel. All my sides are the same length. I do NOT have right angles. What shape am I?

(A) Rectangle   (B) Square

(C) Rhombus   (D) Trapezoid

 # End of Practice Test 1 

Great job finishing the test!

 **My Score**

I got __________ out of 30 questions right.

*Check your answers in the **Answer Key** at the back of the book.*

 *Review any questions you missed. That's how we learn!*

📊 Check Your Score Online!

Visit **ViewMath Academy** to enter your answers and see which topics you need to review. You can also explore lessons, take quizzes, track your scores, and save your progress!

viewmath.com/score/5.1.ND.16

Or go to viewmath.com/score and enter code: 5.1.ND.16

Practice Test 2

📋 *30 Questions*

✏️ Before You Start ✏️

- ✔ **Read each question carefully** before choosing your answer.
- ✔ **Show your work** on scratch paper when you need to.
- ✔ **Skip hard questions** and come back to them later.
- ✔ **Check your answers** when you're done.
- ✔ **Take your time** — there's no rush!

⭐ You've Got This! ⭐

Do your best and show what you know!

1. *What is 5.8×10^2?*

Your Answer:

2. *What is the word form of 0.7?*

(A) Seven hundredths

(B) Seven tenths

(C) Seven thousandths

(D) Seventy tenths

3. *0.999 rounded to the nearest whole number is:*

(A) 0

(B) 1

(C) 0.9

(D) 0.99

4. *Multiply: 56×23*

(A) 1,288

(B) 1,188

(C) 1,278

(D) 280

5. *There are 85 pencils to be shared equally among 6 students. How many pencils will be left over?*

(A) 14

(B) 1

(C) 2

(D) 15

6. *Estimate 78×51 by rounding each factor to the nearest ten.*

(A) 3,500

(B) 4,000

(C) 4,500

(D) 3,000

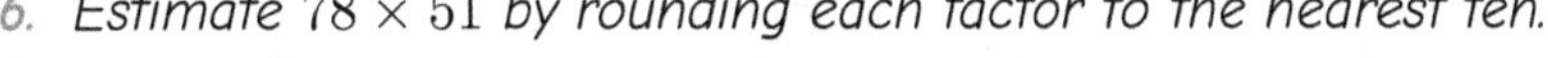

Find more at
ViewMath.com/ND-Grade5

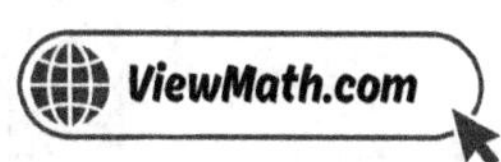

7. What is $10 - 3.47$?

(A) 7.53   (B) 6.53

(C) 6.47   (D) 7.47

8. Each lap around the track is 0.25 miles. Tyler runs 6 laps. How far does he run?

(A) 15.0 miles   (B) 0.15 miles

(C) 1.50 miles   (D) 1.05 miles

9. Divide: $3.2 \div 0.8$

Your Answer:

10. Sara ran $\frac{3}{4}$ of a mile in the morning and $\frac{1}{3}$ of a mile in the afternoon. How far did she run in total?

(A) $\frac{4}{7}$ mile   (B) $1\frac{1}{12}$ miles

(C) $\frac{10}{12}$ mile   (D) $\frac{13}{12}$ miles

11. Estimate $\frac{7}{8} + \frac{4}{5}$.

(A) $\frac{1}{2}$ (B) 1

(C) $1\frac{1}{2}$ (D) 2

12. A pool needs 12 gallons of water. On Monday, $4\frac{5}{8}$ gallons were added. On Tuesday, $3\frac{1}{2}$ gallons were added. How many more gallons are needed?

(A) $3\frac{7}{8}$ gallons   (B) $4\frac{7}{8}$ gallons

(C) $3\frac{1}{8}$ gallons   (D) $4\frac{1}{8}$ gallons

Find more at
ViewMath.com/ND-Grade5

13. Five friends share 8 apples equally. How many apples does each friend get?

(A) $1\frac{2}{5}$ apples

(B) $1\frac{3}{5}$ apples

(C) $\frac{5}{8}$ of an apple

(D) $2\frac{3}{5}$ apples

14. What is $\frac{3}{4} \times 8$?

(A) $\frac{24}{32}$

(B) $\frac{3}{32}$

(C) 6

(D) $\frac{11}{4}$

15. Sam has $\frac{5}{6}$ of a cake. He gives $\frac{1}{5}$ of it to his sister. What fraction of the whole cake did his sister get?

Your Answer:

16. Without computing, is $14 \times \frac{11}{10}$ greater than, less than, or equal to 14?

(A) Greater than 14

(B) Less than 14

(C) Equal to 14

(D) Cannot be determined

17. Convert $3\frac{1}{5}$ to an improper fraction.

Your Answer:

18. Lily has $\frac{7}{8}$ of a gallon of paint. She uses $\frac{1}{2}$ of it. How much paint does she use?

Your Answer:

Find more at
ViewMath.com/ND-Grade5

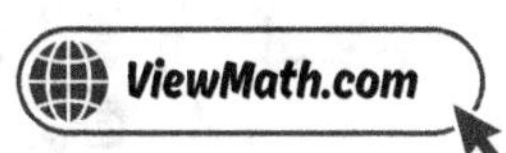

19. Which multiplication equation can you use to check $\frac{1}{3} \div 2 = \frac{1}{6}$?

  (A) $\frac{1}{6} \times 3 = \frac{1}{2}$

  (B) $\frac{1}{6} \times 2 = \frac{1}{3}$

  (C) $\frac{1}{3} \times 2 = \frac{2}{3}$

  (D) $\frac{1}{2} \times 3 = \frac{3}{2}$

20. Evaluate: $2 \times \{5 + [3 \times (1 + 2)]\}$

  (A) 18

  (B) 28

  (C) 22

  (D) 36

21. Which expression means "twice the sum of 10 and 5"?

  (A) $2 + (10 + 5)$

  (B) $2 \times 10 + 5$

  (C) $(10 + 5) \times 2$

  (D) $2 \times (10 - 5)$

22. Rule A: Start at 0, add 4. Rule B: Start at 0, add 12. What is the 3rd ordered pair (A, B)?

  (A) $(4, 12)$

  (B) $(8, 24)$

  (C) $(12, 36)$

  (D) $(0, 0)$

23. Which ordered pair describes a point that is 8 units right and 3 units up from the origin?

  (A) $(3, 8)$

  (B) $(8, 3)$

  (C) $(11, 0)$

  (D) $(0, 11)$

24. A graph shows savings over time. At week 1 the savings are $5, and at week 3 the savings are $15. How much is saved per week?

(A) $3

(B) $5

(C) $10

(D) $15

25. How many pounds are in 10,000 pounds expressed in tons?

(A) 5 T

(B) 10 T

(C) 50 T

(D) 100 T

26. A line plot shows the amounts of paint (in cups) used by 8 students:
$\frac{1}{4}$: 2 X marks $\frac{3}{8}$: 3 X marks $\frac{1}{2}$: 3 X marks
What is the total amount of paint used?

(A) $2\frac{5}{8}$ cups

(B) $3\frac{1}{8}$ cups

(C) $2\frac{7}{8}$ cups

(D) 3 cups

27. Two boxes hold the same number of unit cubes. Does that mean they must be the same shape?

(A) Yes, they must be identical.

(B) No, different shapes can have the same volume.

(C) Yes, if the volumes are equal the shapes are equal.

(D) No, but they must have the same height.

28. A swimming pool is L-shaped. The main section is 12 m × 6 m × 2 m. The smaller section is 4 m × 3 m × 2 m. What is the total volume?

(A) $144\ m^3$

(B) $168\ m^3$

(C) $120\ m^3$

(D) $192\ m^3$

Find more at
ViewMath.com/ND-Grade5

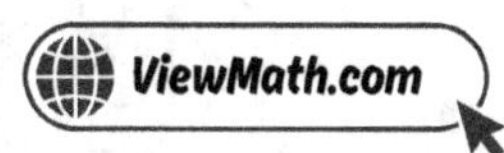

29. Maria packs small cubes with 2-inch edges into a box that is $10 \times 8 \times 6$ inches. How many small cubes fit?

(A) 30

(B) 48

(C) 60

(D) 80

30. A shape has 4 sides and exactly 1 pair of parallel sides. What is it?

(A) Parallelogram

(B) Rectangle

(C) Trapezoid

(D) Rhombus

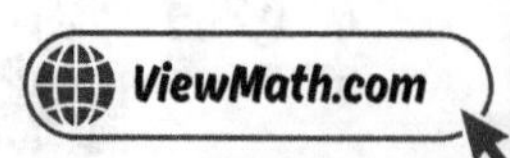

 # End of Practice Test 2 

Great job finishing the test!

☑ My Score

I got ___________ out of 30 questions right.

*Check your answers in the **Answer Key** at the back of the book.*

💡 Review any questions you missed. That's how we learn!

📊 Check Your Score Online!

Visit **ViewMath Academy** to enter your answers and see which topics you need to review. You can also explore lessons, take quizzes, track your scores, and save your progress!

viewmath.com/score/5.1.ND.17

Or go to viewmath.com/score and enter code: 5.1.ND.17

3

Practice Test 3

 *30 Questions*

✏ Before You Start ✏

- ✓ **Read each question carefully** before choosing your answer.
- ✓ **Show your work** on scratch paper when you need to.
- ✓ **Skip hard questions** and come back to them later.
- ✓ **Check your answers** when you're done.
- ✓ **Take your time** — there's no rush!

 ★ You've Got This! ★ 

Do your best and show what you know!

1. What is $700 \div 100$?

  (A) 7,000   (B) 70

  (C) 7   (D) 0.7

2. Write the standard form of "twenty-four thousandths."

  Your Answer:

3. 6.374 rounded to the nearest tenth is:

  (A) 6.4   (B) 6.3

  (C) 6   (D) 6.37

4. Multiply: $5,009 \times 12$

  (A) 60,108   (B) 60,008

  (C) 10,018   (D) 61,108

5. A school is taking 145 students on a field trip. Each van can hold 8 students. How many vans are needed to take all the students?

  (A) 18   (B) 19

  (C) 17   (D) 20

6. Estimate 715×89 by rounding to the nearest hundred and ten.

  Your Answer:

Find more at
ViewMath.com/ND-Grade5

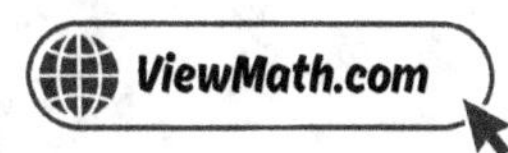

7. *Subtract:* $10 - 4.29$

Your Answer:

8. *What is 0.04×9?*

(A) 3.6

(B) 0.36

(C) 0.036

(D) 36

9. *What is $7.50 \div 0.50$?*

Your Answer:

10. *Add:* $\frac{3}{10} + \frac{1}{4} + \frac{1}{5}$. *Write your answer in simplest form.*

Your Answer:

11. *Estimate $3\frac{7}{8} + 2\frac{1}{5}$.*

(A) 5

(B) 6

(C) 7

(D) $5\frac{1}{2}$

12. *Derek drank $\frac{3}{8}$ of a bottle of water in the morning and $\frac{1}{4}$ in the afternoon. What fraction of the bottle did he drink in all?*

Your Answer:

Find more at
ViewMath.com/ND-Grade5

13. Seven kids share 4 packs of stickers equally. Which expression shows how many packs each kid gets?

(A) $7 \div 4$

(B) 7×4

(C) 4×7

(D) $4 \div 7$

14. What is $9 \times \frac{2}{3}$?

Your Answer:

15. What is $\frac{1}{4} \times \frac{2}{5}$?

Your Answer:

16. Without computing, is $\frac{3}{4} \times 20$ greater than, less than, or equal to 20?

(A) Greater than 20

(B) Less than 20

(C) Equal to 20

(D) Cannot be determined

17. A rectangular garden is $4\frac{1}{2}$ meters long and $2\frac{2}{3}$ meters wide. What is its area?

Your Answer:

18. There are 40 fish in a pond. $\frac{3}{8}$ of them are goldfish. How many goldfish are there?

(A) 10

(B) 12

(C) 15

(D) 24

Find more at
ViewMath.com/ND-Grade5

19. What is $\frac{1}{6} \div 2$?

(A) $\frac{1}{3}$

(B) $\frac{1}{8}$

(C) $\frac{2}{6}$

(D) $\frac{1}{12}$

20. What is the value of $5 \times (12 - 7)$?

(A) 53

(B) 30

(C) 25

(D) 17

21. Write a word phrase for the expression $2 \times (30 - 7)$.

Your Answer:

22. What is the rule for this pattern? $2, 5, 8, 11, 14$

(A) Start at 2, add 3

(B) Start at 2, add 2

(C) Start at 3, add 2

(D) Start at 2, multiply by 3

23. Point R is 3 units right of the origin and 0 units up. What is the ordered pair for R?

(A) $(0, 3)$

(B) $(3, 3)$

(C) $(3, 0)$

(D) $(0, 0)$

24. Rule A: start at 0, add 1. Rule B: start at 0, add 5. What is the value of B when $A = 3$?

(A) 5

(B) 10

(C) 15

(D) 20

Find more at
ViewMath.com/ND-Grade5

25. A recipe calls for 4 cups of milk. How many fluid ounces is that?

(A) 16 fl oz

(B) 24 fl oz

(C) 32 fl oz

(D) 48 fl oz

26. A line plot shows the heights of seedlings (in inches). There are 10 data points. The smallest value is $\frac{1}{8}$ and the largest is $\frac{7}{8}$. What is the range of the data?

(A) $\frac{1}{8}$ inch

(B) $\frac{1}{2}$ inch

(C) $\frac{3}{4}$ inch

(D) 1 inch

27. The bottom layer of a box has 4 cubes along the row and 3 rows. The box is 5 layers tall. What is the volume?

(A) 12 cubic units

(B) 20 cubic units

(C) 35 cubic units

(D) 60 cubic units

28. A U-shaped solid can be thought of as a full block with a rectangular piece removed from the top. The full block is $8 \times 6 \times 5$. The removed piece is $4 \times 2 \times 5$. What is the volume of the U-shape?

(A) 200 cubic units

(B) 240 cubic units

(C) 280 cubic units

(D) 160 cubic units

29. A box has a base area of 42 in^2 and a volume of 252 in^3. What is the height?

(A) 4 in

(B) 5 in

(C) 6 in

(D) 7 in

Find more at
ViewMath.com/ND-Grade5

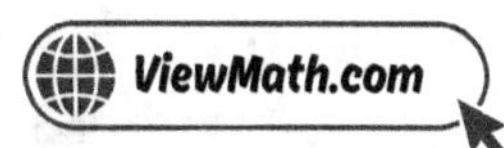

30. *Do all parallelograms have 4 right angles?*

   Ⓐ *Yes, always.*

   Ⓑ *No, only rectangles and squares have 4 right angles.*

   Ⓒ *Yes, because they have 2 pairs of parallel sides.*

   Ⓓ *No, only rhombuses have 4 right angles.*

 *End of Practice Test 3* 

Great job finishing the test!

✅ *My Score*

I got _____________ out of 30 questions right.

*Check your answers in the **Answer Key** at the back of the book.*

💡 *Review any questions you missed. That's how we learn!*

📊 *Check Your Score Online!*

Visit **ViewMath Academy** to enter your answers and see which topics you need to review. You can also explore lessons, take quizzes, track your scores, and save your progress!

viewmath.com/score/5.1.ND.18

Or go to viewmath.com/score and enter code: 5.1.ND.18

Practice Test 4

 30 Questions

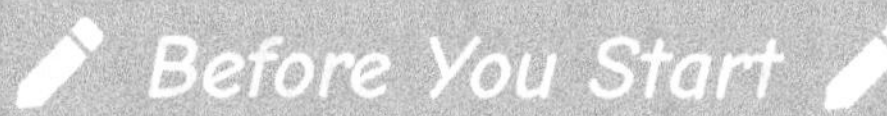

✏️ Before You Start ✏️

- ✔ **Read each question carefully** before choosing your answer.
- ✔ **Show your work** on scratch paper when you need to.
- ✔ **Skip hard questions** and come back to them later.
- ✔ **Check your answers** when you're done.
- ✔ **Take your time** — there's no rush!

  ⭐ You've Got This! ⭐

Do your best and show what you know!

1. Multiplying a number by 10^2 is the same as moving its decimal point:

   (A) 2 places to the left   (B) 2 places to the right

   (C) Adding 2 to the number   (D) Multiplying the number by 2

2. What decimal does $2 \times 10 + 4 \times 1 + 7 \times \frac{1}{1,000}$ equal?

   Your Answer:

3. Ana says 4.35 rounded to the nearest tenth is 4.3. Is she right?

   (A) Yes, because $3 < 5$ (B) No, the correct answer is 4.4

   (C) No, the correct answer is 5   (D) Yes, because the 5 is in tenths

4. Multiply: $3,015 \times 16$

   Your Answer:

5. A delivery truck can carry 45 boxes. There are 300 boxes to deliver. How many trips will the truck need to make to deliver all the boxes?

   (A) 6   (B) 7

   (C) 8   (D) 5

6. Estimate the quotient: $3,150 \div 42$

   (A) 70   (B) 80

   (C) 60   (D) 90

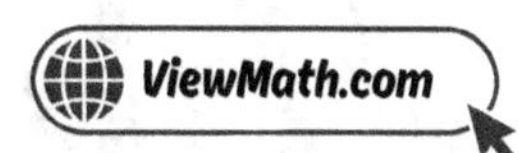

7. *A rope is 8.5 meters long. Marcus cuts off 3.75 meters. How much rope is left?*

   (A) 4.75 *meters*   (B) 5.75 *meters*

   (C) 4.25 *meters*   (D) 5.25 *meters*

8. *What is 0.65×9?*

   Your Answer:

9. *What is $1.20 \div 0.04$?*

   (A) 3   (B) 30

   (C) 300   (D) 0.3

10. *A painter mixed $\frac{3}{8}$ gallon of blue paint with $\frac{1}{2}$ gallon of white paint. How much paint is there in all?*

    Your Answer:

11. *Estimate $5\frac{7}{8} - 2\frac{1}{6}$.*

    Your Answer:

12. *Sophie had $3\frac{1}{4}$ meters of fabric. She used $1\frac{5}{6}$ meters to make a pillow. How much fabric is left?*

    Your Answer:

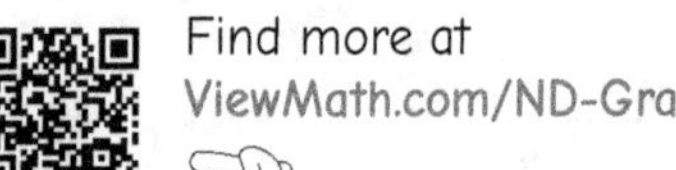

Find more at
ViewMath.com/ND-Grade5

13. Three friends share 2 pies equally. What fraction of a pie does each friend get?

Your Answer:

14. What is $7 \times \frac{1}{2}$ as a mixed number?

(A) $\frac{7}{14}$

(B) 3

(C) $\frac{1}{14}$

(D) $3\frac{1}{2}$

15. What is $\frac{4}{5} \times \frac{5}{8}$ in simplest form?

Your Answer:

16. Multiplying any number by $\frac{6}{6}$ will:

(A) Make it 6 times bigger.

(B) Make it 6 times smaller.

(C) Keep it the same.

(D) Make it equal to 6.

17. What is $1\frac{3}{4} \times \frac{4}{7}$?

(A) $\frac{7}{7}$

(B) 1

(C) $\frac{4}{7}$

(D) $1\frac{3}{7}$

18. A rectangle is $2\frac{1}{4}$ inches long and $1\frac{1}{3}$ inches wide. What is its area?

Your Answer:

Find more at
ViewMath.com/ND-Grade5

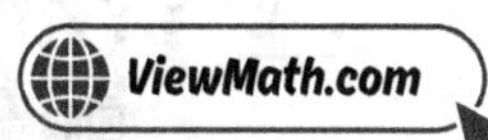

19. A $\frac{1}{6}$-acre garden is divided into 3 equal plots. What is the size of each plot?

(A) $\frac{1}{2}$ acre

(B) $\frac{1}{9}$ acre

(C) $\frac{3}{6}$ acre

(D) $\frac{1}{18}$ acre

20. Where should you place parentheses to make this equation true?

$$6 + 2 \times 4 = 32$$

(A) $6 + (2 \times 4) = 32$

(B) $(6 + 2) \times 4 = 32$

(C) $6 + 2 \times (4) = 32$

(D) No parentheses can make this true

21. Which expression matches "add 12 and 8, then subtract 5"?

(A) $12 + (8 - 5)$

(B) $(12 + 8) - 5$

(C) $12 - (8 + 5)$

(D) $(12 - 8) + 5$

22. Ava earns \$4 per chore. Leo earns \$12 per chore. How many times more does Leo earn than Ava for the same number of chores?

(A) 2 times

(B) 3 times

(C) 4 times

(D) 8 times

23. Which point lies on the y-axis?

(A) $(4, 0)$

(B) $(0, 6)$

(C) $(3, 3)$

(D) $(1, 2)$

Get Online

Find more at
ViewMath.com/ND-Grade5

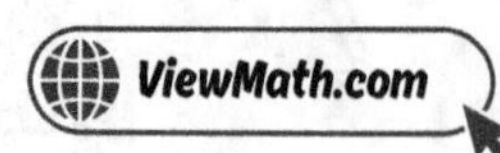

24. A graph shows the points $(2, 6)$, $(4, 12)$, $(6, 18)$. How much does y increase for each increase of 2 in x?

Your Answer:

25. 1 gallon = 4 quarts = 8 pints = 16 cups. How many cups are in 3 pints?

(A) 3 c

(B) 4 c

(C) 6 c

(D) 8 c

26. A line plot shows 5 data points at $\frac{1}{2}$ and 3 data points at $\frac{3}{4}$. What is the combined total of all 8 data points?

Your Answer:

27. A rectangular prism has a bottom layer of 12 unit cubes and is 3 layers tall. What is its volume?

(A) 15 cubic units

(B) 36 cubic units

(C) 24 cubic units

(D) 48 cubic units

28. A combined shape is made of two identical prisms, each $5 \times 3 \times 4$. What is the total volume?

(A) 60 cubic units

(B) 120 cubic units

(C) 180 cubic units

(D) 240 cubic units

29. A toy chest is 4 ft long, 3 ft wide, and 2 ft tall. What is the volume?

(A) 9 ft^3

(B) 12 ft^3

(C) 24 ft^3

(D) 48 ft^3

Find more at
ViewMath.com/ND-Grade5

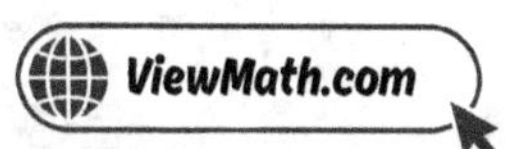

30. *A shape has 4 sides, 4 right angles, and sides of 8 cm and 5 cm. Is it a square or a rectangle?*

Your Answer:

Find more at
ViewMath.com/ND-Grade5

 End of Practice Test 4 

Great job finishing the test!

☑ My Score

I got _____________ out of 30 questions right.

*Check your answers in the **Answer Key** at the back of the book.*

💡 *Review any questions you missed. That's how we learn!*

📊 Check Your Score Online!

Visit **ViewMath Academy** to enter your answers and see which topics you need to review. You can also explore lessons, take quizzes, track your scores, and save your progress!

viewmath.com/score/5.1.ND.19

Or go to viewmath.com/score and enter code: 5.1.ND.19

Practice Test 5

📋 30 Questions

✏️ Before You Start ✏️

- ✔ **Read each question carefully** before choosing your answer.
- ✔ **Show your work** on scratch paper when you need to.
- ✔ **Skip hard questions** and come back to them later.
- ✔ **Check your answers** when you're done.
- ✔ **Take your time** — there's no rush!

⭐ You've Got This! ⭐

Do your best and show what you know!

1. *What is $4{,}200 \div 10^3$?*

Your Answer:

2. $\dfrac{7}{1{,}000}$ *written as a decimal is:*

(A) 0.7 (B) 0.07

(C) 0.007 (D) 7,000

3. *I am a decimal with exactly two decimal places. When rounded to the nearest tenth, I become 6.5. What is the SMALLEST value I could be?*

Your Answer:

4. *Multiply:* $2{,}841 \times 31$

(A) 88,071 (B) 87,071

(C) 88,171 (D) 85,230

5. *Four friends share 15 brownies equally. How many brownies does each friend get?*

(A) 3 (B) 4

(C) $3\frac{3}{4}$ (D) 3 R3

6. *A store sells 68 laptops for $415 each. About how much money did the store make?*

Your Answer:

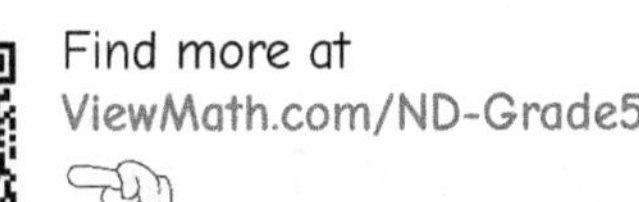
Find more at
ViewMath.com/ND-Grade5

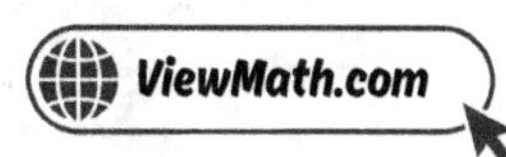

7. *What is* $9.87 - 4.23$*?*

   (A) 5.64

   (B) 5.54

   (C) 4.64

   (D) 5.74

8. *Multiply:* 4.08×7

   (A) 2,856

   (B) 285.6

   (C) 2.856

   (D) 28.56

9. *Divide:* $7.56 \div 0.9$

   (A) 84

   (B) 0.84

   (C) 8.4

   (D) 0.084

10. *What is* $\frac{1}{6} + \frac{2}{3}$*?*

    (A) $\frac{3}{9}$

    (B) $\frac{5}{6}$

    (C) $\frac{3}{6}$

    (D) $\frac{1}{2}$

11. *Round* $\frac{9}{10}$ *to the nearest benchmark* $(0, \frac{1}{2},$ *or* $1)$.

    Your Answer

12. *Lily had 5 cups of juice. She drank* $1\frac{1}{4}$ *cups and gave* $2\frac{1}{3}$ *cups to her friend. How much juice is left?*

    (A) $1\frac{5}{12}$ *cups*

    (B) $2\frac{5}{12}$ *cups*

    (C) $1\frac{7}{12}$ *cups*

    (D) $2\frac{7}{12}$ *cups*

Get Online

Find more at
ViewMath.com/ND-Grade5

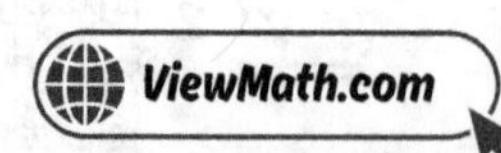

13. A baker has 7 pounds of dough. She divides it equally among 3 loaves. How many pounds of dough is in each loaf?

Your Answer:

14. A garden is 12 meters long. Sarah plants flowers in $\frac{2}{3}$ of the garden. How many meters of flowers does she plant?

(A) 4 meters

(B) 6 meters

(C) 8 meters

(D) 10 meters

15. What is $\frac{3}{5} \times \frac{2}{3}$ in simplest form?

Your Answer:

16. A store sells a toy for $18. During a sale, the price is multiplied by $\frac{2}{3}$. Without computing, is the sale price higher or lower than $18?

(A) Higher, because you are multiplying.

(B) Lower, because $\frac{2}{3} < 1$.

(C) The same, because the price doesn't change.

(D) Higher, because $\frac{2}{3} > 0$.

17. What is $1\frac{1}{2} \times 2\frac{1}{3}$?

(A) $2\frac{1}{6}$

(B) $3\frac{1}{6}$

(C) $3\frac{1}{2}$

(D) $3\frac{5}{6}$

Find more at
ViewMath.com/ND-Grade5

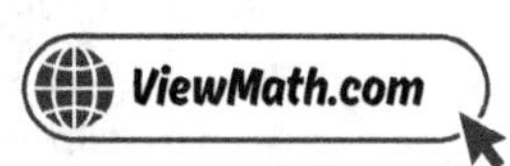

18. Sophia bakes 24 cookies. She gives $\frac{1}{3}$ of them to her neighbor. How many cookies does the neighbor get?

   (A) 3

   (B) 6

   (C) 8

   (D) 12

19. $\frac{1}{2} \div 5$ is the same as:

   (A) $\frac{1}{2} \times 5$

   (B) $5 \div \frac{1}{2}$

   (C) $\frac{1}{2} \times \frac{1}{5}$

   (D) $\frac{5}{2}$

20. Which pair of expressions gives **different** answers?

   (A) $(5+3)+2$ and $5+(3+2)$

   (B) $(7+3) \times 2$ and $7+3 \times 2$

   (C) $(4 \times 2) \times 3$ and $4 \times (2 \times 3)$

   (D) $(6+0) \times 5$ and 6×5

21. Without computing, write $>$, $<$, or $=$ to compare:

$$8 \times (46 + 12) \quad \bigcirc \quad 8 \times 46$$

   Your Answer:

22. What is the rule for this pattern? $0, 5, 10, 15, 20$

   (A) Add 10

   (B) Multiply by 5

   (C) Add 5

   (D) Add 15

23. *Which ordered pair is closest to the origin?*

(A) $(1, 1)$ (B) $(3, 0)$

(C) $(0, 4)$ (D) $(2, 2)$

24. *Runner A runs 2 miles per hour. Runner B runs 5 miles per hour. After 4 hours, how many more miles has Runner B run than Runner A?*

Your Answer:

25. *How many inches are in 5 feet?*

(A) 17 in   (B) 50 in

(C) 60 in   (D) 55 in

26. *A line plot shows snowfall data for 7 days. Six days show $\frac{1}{4}$ inch and one day shows $\frac{3}{4}$ inch. What is the total snowfall?*

(A) $1\frac{3}{4}$ inches   (B) 2 inches

(C) $2\frac{1}{4}$ inches   (D) $2\frac{1}{2}$ inches

27. *What does volume measure?*

(A) The distance around a shape   (B) The flat surface of a shape

(C) The space inside a 3D shape   (D) The length of one edge

Find more at
ViewMath.com/ND-Grade5

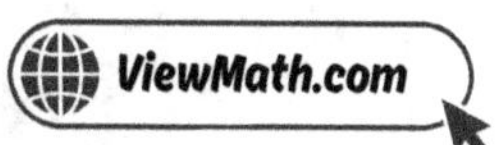

28. A shape is made of a large prism $(9 \times 5 \times 4)$ and a small prism $(3 \times 5 \times 2)$ attached to the top. What is the total volume?

(A) 180 cubic units

(B) 210 cubic units

(C) 30 cubic units

(D) 150 cubic units

29. A cube-shaped storage box has a volume of 216 in^3. What is the length of each edge?

Your Answer:

30. Which of the following is NOT a property of all rectangles?

(A) 4 right angles

(B) Opposite sides are equal

(C) 2 pairs of parallel sides

(D) All 4 sides are equal

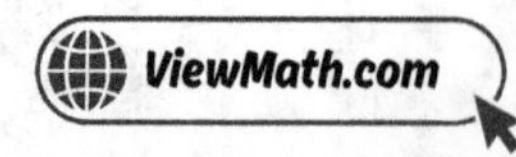

⭐ End of Practice Test 5 ⭐

Great job finishing the test!

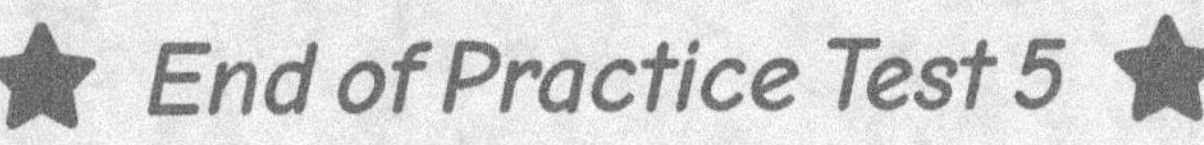 My Score

I got _____________ out of 30 questions right.

*Check your answers in the **Answer Key** at the back of the book.*

💡 *Review any questions you missed. That's how we learn!*

📊 Check Your Score Online!

Visit **ViewMath Academy** to enter your answers and see which topics you need to review. You can also explore lessons, take quizzes, track your scores, and save your progress!

viewmath.com/score/5.1.ND.20

Or go to viewmath.com/score and enter code: 5.1.ND.20

6

Practice Test 6

 30 Questions

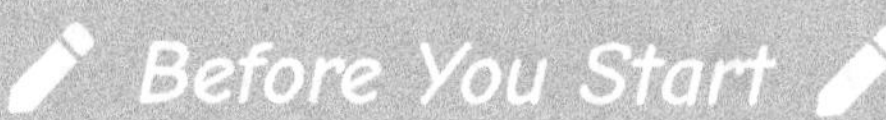

✏ Before You Start ✏

- ✓ **Read each question carefully** before choosing your answer.
- ✓ **Show your work** on scratch paper when you need to.
- ✓ **Skip hard questions** and come back to them later.
- ✓ **Check your answers** when you're done.
- ✓ **Take your time** — there's no rush!

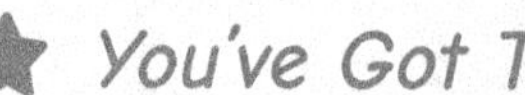  You've Got This! ⭐

Do your best and show what you know!

1. Which of the following equals 0.7×10^3?

   (A) 7

   (B) 70

   (C) 700

   (D) 7,000

2. Write the word form of 5.3.

   Your Answer:

3. 9.96 rounded to the nearest tenth is:

   (A) 9.9

   (B) 9.10

   (C) 10.0

   (D) 10.1

4. What is the product of 73×99?

   (A) 7,227

   (B) 7,327

   (C) 7,127

   (D) 657

5. A teacher has 75 stickers to give equally to 22 students. How many stickers does each student get?

   (A) 3

   (B) 4

   (C) 9

   (D) $3\frac{9}{22}$

6. Estimate 83×21 by rounding to the nearest ten.

   Your Answer:

Find more at
ViewMath.com/ND-Grade5

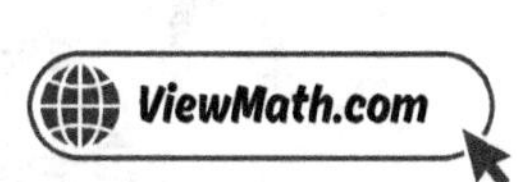

7. *Subtract:* 8.45 − 5.45

Your Answer:

8. *What is* 0.06×4?

(A) 2.4

(B) 0.24

(C) 0.024

(D) 24

9. *What is* $3.6 \div 0.4$?

(A) 0.9

(B) 90

(C) 9

(D) 0.09

10. *What is* $\frac{3}{10} + \frac{2}{5}$?

(A) $\frac{5}{15}$

(B) $\frac{5}{10}$

(C) $\frac{7}{10}$

(D) $\frac{1}{2}$

11. *Without computing, which sum is greater:* $\frac{3}{4} + \frac{7}{8}$ *or* $\frac{2}{5} + \frac{1}{3}$?

(A) $\frac{2}{5} + \frac{1}{3}$

(B) *They are equal.*

(C) $\frac{3}{4} + \frac{7}{8}$

(D) *Cannot be determined without computing.*

12. *Jake ran* $\frac{3}{4}$ *of a mile. Emma ran* $\frac{2}{3}$ *of a mile. How much farther did Jake run than Emma?*

(A) $\frac{1}{1}$ *mile*

(B) $\frac{1}{12}$ *mile*

(C) $\frac{5}{12}$ *mile*

(D) $\frac{17}{12}$ *miles*

Find more at
ViewMath.com/ND-Grade5

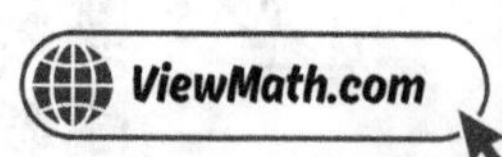

13. A teacher has 9 feet of ribbon. She cuts it into 4 equal pieces. How long is each piece?

(A) $2\frac{1}{4}$ feet

(B) $2\frac{1}{2}$ feet

(C) $\frac{4}{9}$ feet

(D) $4\frac{1}{2}$ feet

14. Ben says $\frac{2}{7} \times 5 = \frac{10}{35}$. What mistake did he make?

(A) He multiplied the numerator wrong.

(B) He multiplied both the numerator and the denominator by 5.

(C) He forgot to simplify.

(D) He switched the numerator and denominator.

15. What is $\frac{5}{6} \times \frac{3}{5}$ in simplest form?

(A) $\frac{15}{30}$

(B) $\frac{1}{2}$

(C) $\frac{8}{11}$

(D) $\frac{3}{6}$

16. Without computing, is $\frac{6}{6} \times 45$ greater than, less than, or equal to 45?

(A) Greater than 45

(B) Less than 45

(C) Equal to 45

(D) Cannot be determined

17. What is $1\frac{1}{4} \times 1\frac{1}{3}$?

(A) $1\frac{1}{12}$

(B) $1\frac{2}{3}$

(C) $1\frac{7}{12}$

(D) $2\frac{1}{12}$

18. A ribbon is $4\frac{1}{2}$ feet long. Jenna uses $\frac{2}{3}$ of it. How much ribbon does she use?

(A) 2 feet

(B) 3 feet

(C) $3\frac{1}{2}$ feet

(D) $6\frac{3}{4}$ feet

19. A pizza shop has $\frac{1}{4}$ of a pizza left. Three customers share it equally. How much does each customer get?

(A) $\frac{3}{4}$ of a pizza

(B) $\frac{1}{7}$ of a pizza

(C) $\frac{1}{12}$ of a pizza

(D) $\frac{4}{3}$ of a pizza

20. Is the following statement true or false?

$$(8 + 2) \times 3 = 8 + 2 \times 3$$

(A) True, both equal 30

(B) True, both equal 14

(C) False, the left side is 30 and the right side is 14

(D) False, the left side is 14 and the right side is 30

21. Which expression means "add 8 and 7, then multiply by 2"?

(A) $8 + 7 \times 2$

(B) $2 \times (8 + 7)$

(C) $(8 \times 7) + 2$

(D) $2 + (8 \times 7)$

22. Two patterns both start at 0. Pattern A: add 6. Pattern B: add 18. What is the 5th term in Pattern B?

Your Answer:

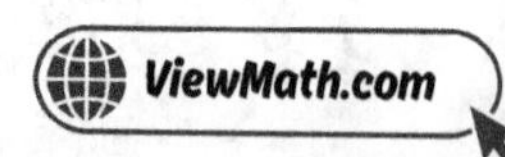

23. A point has x-coordinate 0 and y-coordinate 8. What is the ordered pair?

(A) $(8, 0)$

(B) $(0, 8)$

(C) $(8, 8)$

(D) $(0, 0)$

24. Two friends track their reading. Amy reads 10 pages per day. Ben reads 5 pages per day. After 4 days, how many more pages has Amy read than Ben?

(A) 5

(B) 10

(C) 15

(D) 20

25. A punch recipe makes 2 gallons. How many cups is that?

Your Answer:

26. A line plot shows rainfall data (in inches):

$\frac{1}{8}$: 2 X marks $\frac{1}{4}$: 4 X marks $\frac{3}{8}$: 1 X mark

What is the difference between the greatest and least rainfall values on the plot?

(A) $\frac{1}{8}$ inch

(B) $\frac{1}{4}$ inch

(C) $\frac{3}{8}$ inch

(D) $\frac{1}{2}$ inch

27. A rectangular prism is 5 units long, 5 units wide, and 5 units tall. What is its volume?

Your Answer:

Find more at
ViewMath.com/ND-Grade5

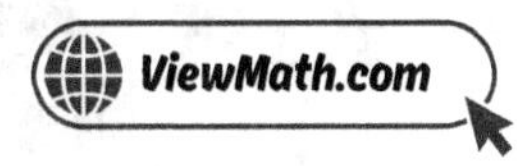

28. A combined shape is split into Prism A $(7 \times 3 \times 5)$ and Prism B $(4 \times 3 \times 5)$. What is the total volume?

   (A) 105 *cubic units*

   (B) 60 *cubic units*

   (C) 165 *cubic units*

   (D) 210 *cubic units*

29. A bin is 12 in long, 8 in wide, and 10 in tall. It is filled to $\frac{3}{4}$ of its height with sand. How many cubic inches of sand are in the bin?

   Your Answer:

30. Which statement is always true?

   (A) All rectangles have 4 equal sides.

   (B) All parallelograms have right angles.

   (C) All rhombuses have 4 equal sides.

   (D) All trapezoids have 2 pairs of parallel sides.

Find more at
ViewMath.com/ND-Grade5

⭐ End of Practice Test 6 ⭐

Great job finishing the test!

 My Score

I got _____________ out of 30 questions right.

*Check your answers in the **Answer Key** at the back of the book.*

💡 *Review any questions you missed. That's how we learn!*

📊 Check Your Score Online!

Visit **ViewMath Academy** to enter your answers and see which topics you need to review. You can also explore lessons, take quizzes, track your scores, and save your progress!

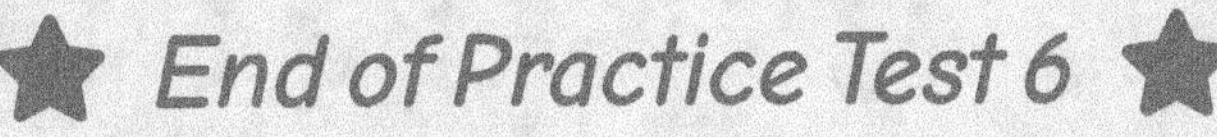

viewmath.com/score/5.1.ND.21

Or go to viewmath.com/score and enter code: 5.1.ND.21

Practice Test 7

 30 Questions

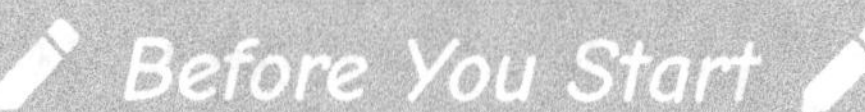

✏ Before You Start ✏

- ✔ **Read each question carefully** before choosing your answer.
- ✔ **Show your work** on scratch paper when you need to.
- ✔ **Skip hard questions** and come back to them later.
- ✔ **Check your answers** when you're done.
- ✔ **Take your time** — there's no rush!

 You've Got This! 

Do your best and show what you know!

1. Find the missing number. $? \times 10^3 = 7{,}000$

Your Answer:

2. What is the standard form of $4 \times 10 + 8 \times 1 + 3 \times \frac{1}{100}$?

(A) 48.03

(B) 4,803

(C) 48.3

(D) 4.83

3. 9.9952 rounded to the nearest tenth is:

(A) 9.9

(B) 10.0

(C) 9.99

(D) 10.1

4. A library has 24 bookshelves. Each bookshelf holds 350 books. How many books are in the library?

(A) 8,400

(B) 8,200

(C) 7,400

(D) 8,600

5. Sarah has $50 to buy notebooks. Each notebook costs $6. How many notebooks can she buy?

(A) 8

(B) 9

(C) 7

(D) 10

6. Estimate $5{,}392 \div 88$ using compatible numbers.

Your Answer:

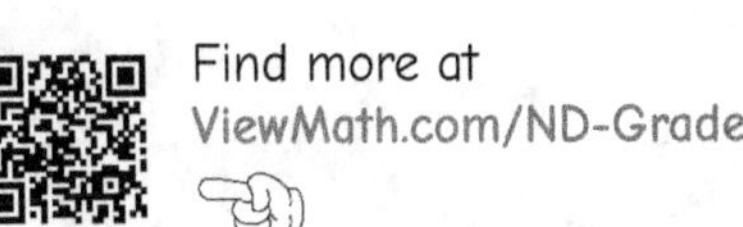
Find more at
ViewMath.com/ND-Grade5

7. *Ava had $20.00 and bought a book for $12.49. How much change did she get?*

   (A) $8.51

   (B) $7.51

   (C) $7.49

   (D) $8.49

8. *What is 1.08×5?*

   Your Answer:

9. *Dividing by 0.1 gives the same result as:*

   (A) Dividing by 10

   (B) Multiplying by 10

   (C) Multiplying by 0.1

   (D) Subtracting 0.1

10. *What is $\frac{3}{5} + \frac{1}{4}$?*

    (A) $\frac{4}{9}$

    (B) $\frac{4}{20}$

    (C) $\frac{17}{20}$

    (D) $\frac{7}{20}$

11. *Estimate $6\frac{1}{9} - 2\frac{4}{5}$.*

    Your Answer:

12. *A carpenter has a plank that is $5\frac{1}{2}$ feet long. He cuts off $2\frac{3}{4}$ feet. How much is left?*

    Your Answer:

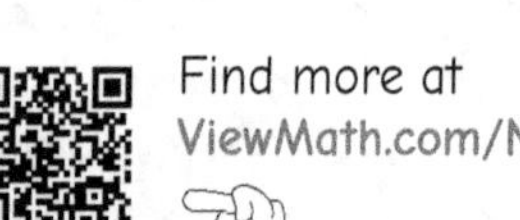

Find more at
ViewMath.com/ND-Grade5

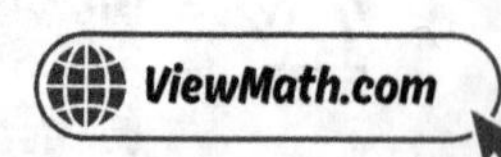

13. What does the fraction $\frac{3}{4}$ mean as a division problem?

(A) $4 \div 3$

(B) $3 \div 4$

(C) 3×4

(D) $4 - 3$

14. A class has 35 students. $\frac{3}{7}$ of them brought their lunch from home. How many students brought their lunch?

Your Answer:

15. Amy says $\frac{1}{4} \times \frac{1}{3} = \frac{1}{7}$. What mistake did she make?

(A) She multiplied the numerators wrong.

(B) She added the denominators instead of multiplying them.

(C) She forgot to simplify.

(D) She switched the numerator and denominator.

16. List the following products from least to greatest without computing: $40 \times \frac{1}{4}$, $40 \times \frac{3}{4}$, $40 \times \frac{5}{4}$.

Your Answer:

17. What is $3\frac{1}{3} \times 2\frac{1}{4}$? Write your answer as a mixed number.

Your Answer:

18. A class has 30 students. $\frac{2}{5}$ brought their lunch from home. How many students brought their lunch?

(A) 6

(B) 10

(C) 12

(D) 15

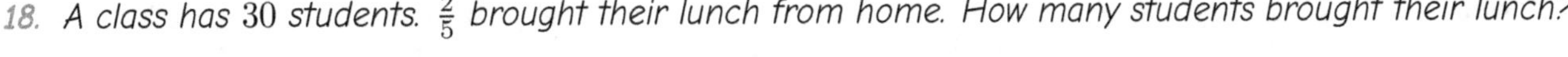

Find more at
ViewMath.com/ND-Grade5

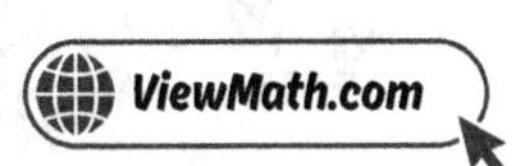

19. Mom has $\frac{1}{2}$ of a cake. She splits it equally among 4 children. What fraction of the whole cake does each child get?

   (A) $\frac{1}{2}$

   (B) $\frac{1}{4}$

   (C) $\frac{1}{6}$

   (D) $\frac{1}{8}$

20. What is the **first step** when evaluating $4 \times [3 + (10 - 6)]$?

   (A) Multiply 4×3

   (B) Add $3 + 10$

   (C) Subtract $10 - 6$

   (D) Multiply 4×10

21. Which is an **expression** (not an equation)?

   (A) $4 + 5 = 9$

   (B) $(3 \times 2) + 1$

   (C) $10 - 3 = 7$

   (D) $6 \times 2 = 12$

22. Pattern A: $0, 1, 2, 3, 4$. Pattern B: $0, 4, 8, 12, 16$. What is the 5th ordered pair (A, B)?

   (A) $(4, 16)$

   (B) $(5, 20)$

   (C) $(4, 12)$

   (D) $(3, 16)$

23. The x-axis is the _________ number line.

   (A) Vertical

   (B) Diagonal

   (C) Horizontal

   (D) Curved

Find more at
ViewMath.com/ND-Grade5

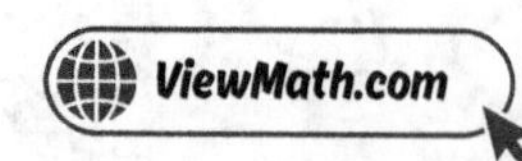

24. A cyclist rides 4 miles every hour. Which point does **NOT** belong on the graph of this relationship (hours, miles)?

(A) $(1, 4)$

(B) $(2, 8)$

(C) $(3, 10)$

(D) $(4, 16)$

25. A football field is 100 yards long. How many feet is that?

(A) 200 ft

(B) 300 ft

(C) 400 ft

(D) 1,200 ft

26. A scientist records daily rainfall (in inches): $0, \frac{1}{4}, \frac{1}{4}, \frac{1}{2}, \frac{3}{4}, 0, \frac{1}{2}$. What is the total rainfall for the week?

Your Answer:

27. A cube is 4 units long, 4 units wide, and 4 units tall. What is its volume?

(A) 12 cubic units

(B) 16 cubic units

(C) 48 cubic units

(D) 64 cubic units

28. A solid is made of two prisms stacked on top of each other. Bottom: $8 \times 6 \times 4$. Top: $4 \times 6 \times 3$. What is the total volume?

(A) 192 cubic units

(B) 72 cubic units

(C) 264 cubic units

(D) 240 cubic units

29. A planter box is 18 in long, 12 in wide, and 8 in deep. How much soil does it hold?

(A) $38 \ in^3$

(B) $216 \ in^3$

(C) $1,728 \ in^3$

(D) $1,296 \ in^3$

Find more at
ViewMath.com/ND-Grade5

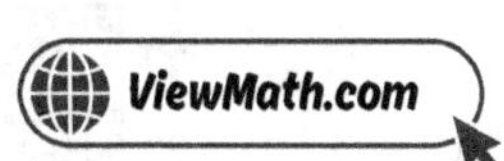

30. Which statement is sometimes true but NOT always true?

(A) All rectangles have 4 right angles.   (B) A parallelogram has right angles.

(C) All squares have 4 equal sides.   (D) All rhombuses have 4 equal sides.

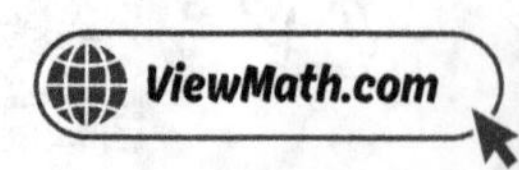

★ *End of Practice Test 7* ★

Great job finishing the test!

 **My Score**

I got __________ out of 30 questions right.

*Check your answers in the **Answer Key** at the back of the book.*

💡 *Review any questions you missed. That's how we learn!*

📊 **Check Your Score Online!**

Visit **ViewMath Academy** to enter your answers and see which topics you need to review. You can also explore lessons, take quizzes, track your scores, and save your progress!

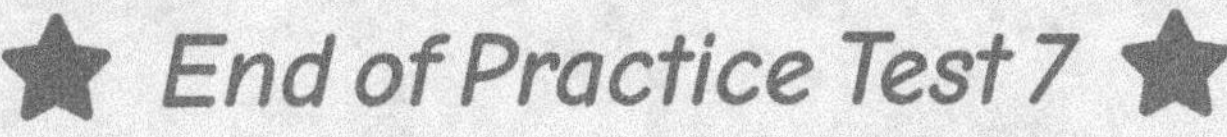

viewmath.com/score/5.1.ND.22

Or go to viewmath.com/score and enter code: 5.1.ND.22

8

Practice Test 8

 30 Questions

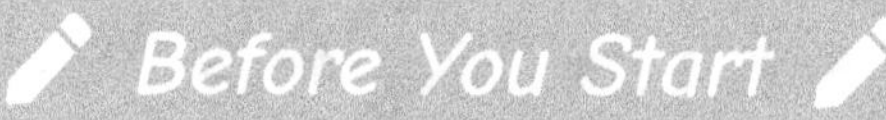

✏️ Before You Start ✏️

- ✔ **Read each question carefully** before choosing your answer.
- ✔ **Show your work** on scratch paper when you need to.
- ✔ **Skip hard questions** and come back to them later.
- ✔ **Check your answers** when you're done.
- ✔ **Take your time** — there's no rush!

 You've Got This! 

Do your best and show what you know!

1. Which exponent makes this equation true? $0.006 \times 10^{?} = 60$

(A) 10^2 (B) 10^3

(C) 10^4 (D) 10^5

2. Which number is correctly expressed as "two hundred and five hundredths"?

(A) 2,005   (B) 200.5

(C) 200.05   (D) 2.005

3. 2.999 rounded to the nearest hundredth is ____.

Your Answer:

4. When multiplying 48×36 using the standard algorithm, what is the second partial product (the result of multiplying by the tens digit)?

(A) 144   (B) 1,440

(C) 288   (D) 1,728

5. A factory produces 530 toys. They are packed into crates that hold 24 toys each. How many toys will be in the partially filled crate?

(A) 22   (B) 2

(C) 23   (D) 21

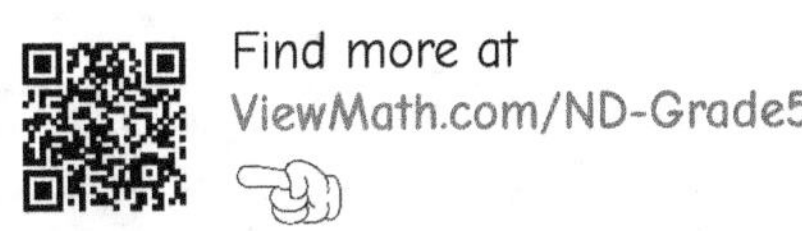

6. *Estimate the quotient:* $5,520 \div 68$

(A) 70

(B) 80

(C) 90

(D) 60

7. *Maria had $50.00. She bought shoes for $32.99. How much does she have left?*

(A) $17.01

(B) $18.01

(C) $17.99

(D) $27.01

8. *Multiply:* 0.25×6

Your Answer:

9. *When you divide a number by a decimal less than 1, the quotient is:*

(A) Smaller than the dividend

(B) Equal to the dividend

(C) Larger than the dividend

(D) Always a whole number

10. *What is the LCD of 4 and 10?*

(A) 40

(B) 20

(C) 10

(D) 14

11. *Estimate* $\frac{5}{6} - \frac{2}{9}$.

(A) 0

(B) $\frac{1}{2}$

(C) 1

(D) $1\frac{1}{2}$

Find more at
ViewMath.com/ND-Grade5

12. A baker used $3\frac{1}{2}$ cups of flour in the morning and $2\frac{3}{4}$ cups in the afternoon. How much flour did the baker use in all?

(A) $5\frac{4}{6}$ cups

(B) $6\frac{1}{4}$ cups

(C) $5\frac{1}{4}$ cups

(D) $6\frac{3}{4}$ cups

13. Three children share 1 sandwich equally. What fraction of the sandwich does each child get?

(A) $\frac{3}{1}$

(B) $\frac{1}{3}$

(C) 3

(D) $\frac{1}{2}$

14. What is $\frac{5}{8} \times 4$?

(A) $\frac{5}{2}$

(B) $\frac{5}{32}$

(C) $2\frac{1}{2}$

(D) $\frac{9}{8}$

15. What is $\frac{2}{3} \times \frac{4}{5}$?

(A) $\frac{6}{8}$

(B) $\frac{8}{15}$

(C) $\frac{6}{15}$

(D) $\frac{8}{8}$

16. Without computing, is $\frac{7}{3} \times 9$ greater than, less than, or equal to 9?

Your Answer:

17. What is $\frac{3}{4} \times 2\frac{2}{5}$?

(A) $\frac{9}{10}$

(B) $1\frac{4}{5}$

(C) $2\frac{3}{20}$

(D) $\frac{6}{20}$

Find more at
ViewMath.com/ND-Grade5

ViewMath.com

18. Jake ate $\frac{1}{4}$ of a 12-slice pizza. Which equation shows how many slices he ate?

   (A) $12 \div \frac{1}{4} = 48$

   (B) $12 - \frac{1}{4} = 11\frac{3}{4}$

   (C) $12 + \frac{1}{4} = 12\frac{1}{4}$

   (D) $\frac{1}{4} \times 12 = 3$

19. What is $\frac{1}{5} \div 2$?

   (A) $\frac{2}{5}$

   (B) $\frac{1}{3}$

   (C) $\frac{1}{10}$

   (D) $\frac{1}{7}$

20. Which expression does **NOT** equal 20?

   (A) $(10 + 10)$

   (B) $4 \times (2 + 3)$

   (C) $(15 - 5) \times 2$

   (D) $2 + (3 \times 5)$

21. Without computing, which expression is larger?

   (A) 7×325

   (B) $7 \times (325 + 18)$

   (C) They are equal

   (D) Cannot be determined

22. If each term in Pattern B is 5 times the corresponding term in Pattern A, and the 3rd term of A is 6, what is the 3rd term of B?

   (A) 11

   (B) 25

   (C) 30

   (D) 35

23. The y-coordinate tells you how far to move:

(A) Right

(B) Left

(C) Up

(D) Down

24. On a graph, a steeper line means:

(A) A slower rate of change

(B) A faster rate of change

(C) No change at all

(D) The values are decreasing

25. A truck weighs 3 tons. A trailer weighs 4,000 pounds. How many pounds do they weigh together?

Your Answer:

26. A line plot shows the weights of apples (in pounds). The data points are:
$\frac{1}{4}, \frac{1}{4}, \frac{3}{8}, \frac{1}{2}, \frac{3}{8}, \frac{1}{4}$

What is the difference between the heaviest and lightest apple?

(A) $\frac{1}{8}$ lb

(B) $\frac{1}{4}$ lb

(C) $\frac{3}{8}$ lb

(D) $\frac{1}{2}$ lb

27. Noah builds a rectangular prism using 48 unit cubes. The bottom layer is 6 cubes by 4 cubes. How many layers tall is his prism?

Your Answer:

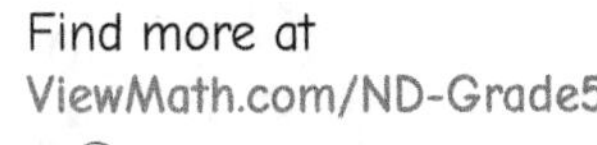

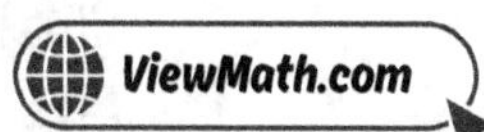

28. *Why is volume called "additive"?*

(A) *You always add the dimensions together.*

(B) *You can split a shape into parts, find each volume, and add them.*

(C) *You add the areas of all the faces.*

(D) *You add length plus width plus height.*

29. *A swimming pool is 25 m long, 10 m wide, and 2 m deep. How many cubic meters of water does it hold?*

(A) $37\ m^3$

(B) $250\ m^3$

(C) $500\ m^3$

(D) $5,000\ m^3$

30. *"A rectangle is a square" — is this always, sometimes, or never true?*

(A) *Always true*

(B) *Sometimes true*

(C) *Never true*

(D) *Cannot be determined*

 # ★ End of Practice Test 8 ★ 

Great job finishing the test!

☑ My Score

I got __________ out of 30 questions right.

*Check your answers in the **Answer Key** at the back of the book.*

💡 Review any questions you missed. That's how we learn!

📊 Check Your Score Online!

Visit **ViewMath Academy** to enter your answers and see which topics you need to review. You can also explore lessons, take quizzes, track your scores, and save your progress!

viewmath.com/score/5.1.ND.23

Or go to viewmath.com/score and enter code: 5.1.ND.23

Practice Test 9

 30 Questions

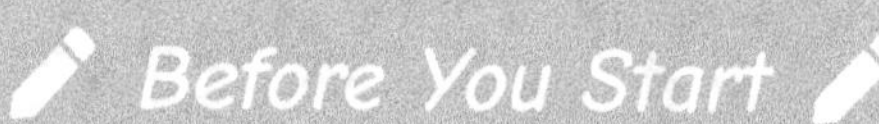

✏ Before You Start ✏

- ✔ **Read each question carefully** before choosing your answer.
- ✔ **Show your work** on scratch paper when you need to.
- ✔ **Skip hard questions** and come back to them later.
- ✔ **Check your answers** when you're done.
- ✔ **Take your time** — there's no rush!

 You've Got This! 

Do your best and show what you know!

1. *What is* 23×10^2?

   (A) 230

   (B) 2,300

   (C) 23,000

   (D) 46

2. *Which decimal equals* $6 \times 100 + 9 \times 1 + 5 \times \frac{1}{100} + 3 \times \frac{1}{1,000}$?

   (A) 609.53

   (B) 609.053

   (C) 6,090.53

   (D) 60.953

3. *46.875 rounded to the nearest hundredth is:*

   (A) 46.8

   (B) 46.87

   (C) 46.88

   (D) 47

4. *What is the product of* $3,050 \times 24$?

   (A) 73,200

   (B) 72,200

   (C) 12,200

   (D) 74,200

5. *A rope is 100 feet long. It is cut into pieces that are exactly 9 feet long. How many feet of rope are left over?*

   (A) 11

   (B) 1

   (C) 12

   (D) 2

6. *Estimate* 47×63 *by rounding each factor to the nearest ten.*

   (A) 2,500

   (B) 3,000

   (C) 2,961

   (D) 3,500

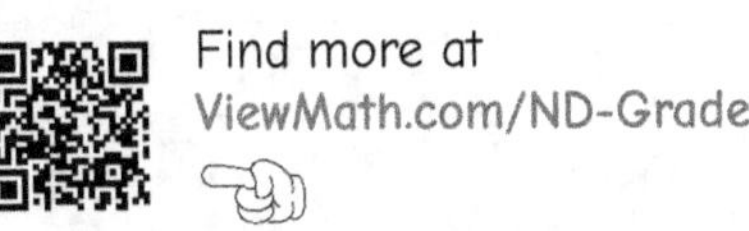

Find more at
ViewMath.com/ND-Grade5

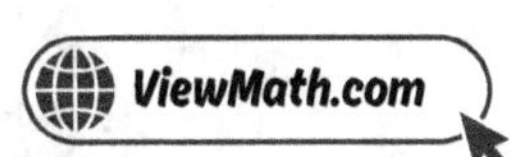

7. *Subtract:* $6 - 2.45$

(A) 4.55   (B) 3.55

(C) 3.45   (D) 4.45

8. *Multiply:* 6.15×2

(A) 12.3   (B) 1.23

(C) 12.30   (D) 123

9. *Divide:* $8.12 \div 0.4$

Your Answer:

10. *What is* $\frac{2}{3} + \frac{5}{9}$?

(A) $\frac{7}{9}$ (B) $\frac{7}{12}$

(C) $1\frac{2}{9}$ (D) $\frac{11}{9}$

11. *A student computed* $\frac{7}{8} + \frac{4}{5} = \frac{11}{13}$. *Use estimation to decide if this is reasonable.*

(A) *Yes, because* $\frac{11}{13}$ *is close to 1.*   (B) *No, because the estimate is about 2 but* $\frac{11}{13} < 1$.

(C) *Yes, because both fractions are less than 1.*   (D) *No, because* $11 + 13 = 24$.

12. *Maria practiced piano for* $1\frac{1}{2}$ *hours on Monday and* $2\frac{1}{3}$ *hours on Tuesday. How many hours did she practice in total?*

(A) $3\frac{2}{5}$ *hours*   (B) $3\frac{5}{6}$ *hours*

(C) $4\frac{5}{6}$ *hours*   (D) $3\frac{1}{6}$ *hours*

Find more at
ViewMath.com/ND-Grade5

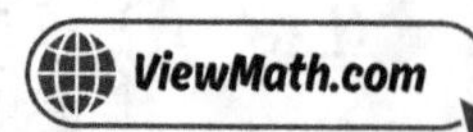

13. Between which two whole numbers does $\frac{11}{4}$ lie?

(A) 1 and 2

(B) 2 and 3

(C) 3 and 4

(D) 4 and 5

14. What is $\frac{5}{6} \times 3$ in simplest form?

(A) $\frac{15}{18}$

(B) $\frac{15}{6}$

(C) $2\frac{1}{2}$

(D) $2\frac{1}{6}$

15. A banner is $\frac{2}{3}$ foot wide and $\frac{5}{6}$ foot tall. What is its area?

(A) $\frac{7}{9}$ square foot

(B) $\frac{5}{9}$ square foot

(C) $\frac{10}{9}$ square foot

(D) $\frac{7}{18}$ square foot

16. A recipe uses $\frac{3}{4}$ cup of flour. Emma wants to make $\frac{2}{3}$ of the recipe. Will she use more or less than $\frac{3}{4}$ cup?

(A) More, because she is multiplying.

(B) Less, because $\frac{2}{3} < 1$.

(C) The same amount.

(D) More, because $\frac{2}{3} > \frac{1}{2}$.

17. What is $2\frac{2}{3} \times 1\frac{1}{8}$?

(A) $2\frac{3}{4}$

(B) 3

(C) $3\frac{1}{4}$

(D) $2\frac{3}{24}$

18. A trail is 6 miles long. Sam hikes $\frac{5}{6}$ of the trail. How far does he hike?

(A) 4 miles

(B) 5 miles

(C) $\frac{5}{6}$ miles

(D) $\frac{30}{6}$ miles

19. Find the missing number: $\frac{1}{4} \div ? = \frac{1}{12}$.

Your Answer

20. Evaluate: $3 \times [8 - (2 + 1)]$

(A) 21

(B) 19

(C) 15

(D) 27

21. Elena has \$20. She buys 3 notebooks for \$4 each. Which expression shows how much money she has left?

(A) $20 - 3 + 4$

(B) $20 - (3 \times 4)$

(C) $(20 - 3) \times 4$

(D) $20 \times 3 - 4$

22. Rule A: Start at 0, add 2. Rule B: Start at 0, add 10. What is the 4th ordered pair (A, B)?

(A) $(4, 20)$

(B) $(6, 30)$

(C) $(8, 40)$

(D) $(2, 10)$

23. To plot the point $(4, 7)$, you start at the origin and move:

(A) 4 units up, 7 units right

(B) 7 units right, 4 units up

(C) 4 units right, 7 units up

(D) 4 units left, 7 units down

24. A plant grows 2 cm per day. The ordered pairs are (day, height). What is the plant's height after 7 days?

(A) 9 cm

(B) 12 cm

(C) 14 cm

(D) 16 cm

25. Sarah needs 6 quarts of juice. The store sells juice in pints. How many pints does she need?

(A) 3 pt

(B) 6 pt

(C) 12 pt

(D) 24 pt

26. A line plot shows plant growth data. The values on the number line go from $\frac{1}{8}$ to $\frac{1}{2}$. What fractions should the tick marks show?

(A) $\frac{1}{8}, \frac{1}{4}, \frac{1}{2}$

(B) $\frac{1}{8}, \frac{2}{8}, \frac{3}{8}, \frac{4}{8}$

(C) $\frac{1}{8}, \frac{1}{6}, \frac{1}{4}, \frac{1}{2}$

(D) $\frac{1}{4}, \frac{1}{2}, \frac{3}{4}, 1$

27. A prism has a bottom layer of 8 cubes and a total volume of 40 cubic units. How many layers tall is it?

(A) 4

(B) 5

(C) 8

(D) 32

28. A block is $10 \times 10 \times 6$ cm. A $5 \times 5 \times 6$ cm piece is removed from one corner. What is the remaining volume?

(A) $450 \ cm^3$

(B) $500 \ cm^3$

(C) $550 \ cm^3$

(D) $600 \ cm^3$

Find more at
ViewMath.com/ND-Grade5

29. A rectangular garden bed has a volume of 96 ft³. It is 8 ft long and 2 ft deep. How wide is it?

(A) 4 ft

(B) 6 ft

(C) 8 ft

(D) 12 ft

30. Which shape is both a rectangle AND a rhombus?

(A) Parallelogram

(B) Trapezoid

(C) Square

(D) Quadrilateral

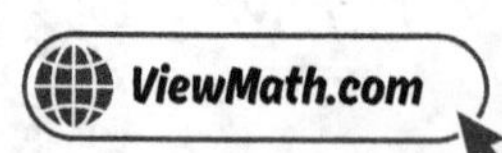

★ End of Practice Test 9 ★

Great job finishing the test!

 My Score

I got ___________ out of 30 questions right.

*Check your answers in the **Answer Key** at the back of the book.*

💡 Review any questions you missed. That's how we learn!

📊 Check Your Score Online!

Visit **ViewMath Academy** to enter your answers and see which topics you need to review. You can also explore lessons, take quizzes, track your scores, and save your progress!

viewmath.com/score/5.1.ND.24

Or go to viewmath.com/score and enter code: 5.1.ND.24

Practice Test 10

 30 Questions

✏ Before You Start ✏

- ✔ **Read each question carefully** before choosing your answer.
- ✔ **Show your work** on scratch paper when you need to.
- ✔ **Skip hard questions** and come back to them later.
- ✔ **Check your answers** when you're done.
- ✔ **Take your time** — there's no rush!

 ★ You've Got This! ★ 

Do your best and show what you know!

1. Find the missing number: $? \div 10^2 = 3.7$

(A) 0.037

(B) 37

(C) 370

(D) 3,700

2. "Eight hundred and twelve thousandths" written as a decimal is:

(A) 800.12

(B) 800.012

(C) 812.000

(D) 0.812

3. A bottle contains 1.655 liters. Rounded to the nearest tenth, how many liters does it hold?

Your Answer:

4. What is 415×38?

(A) 15,770

(B) 14,770

(C) 15,670

(D) 16,770

5. A baker makes 350 cookies and packs them into bags of 24. How many cookies are in the partially filled bag?

(A) 14

(B) 15

(C) 16

(D) 13

6. A school buys 88 boxes of markers. Each box has 115 markers. Which is the best estimate for the total number of markers?

(A) 9,000

(B) 10,000

(C) 8,000

(D) 11,000

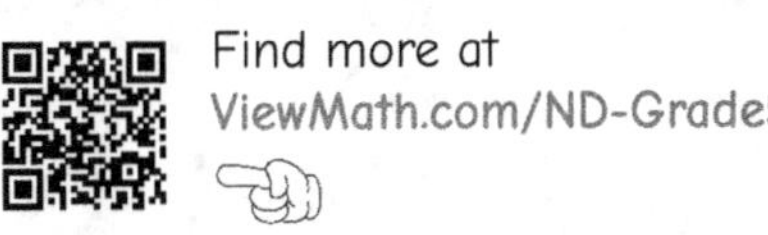
Find more at
ViewMath.com/ND-Grade5

7. When subtracting $4 - 1.7$, what should you write 4 as?

(A) 4.7

(B) 40

(C) 4.0

(D) 0.4

8. Which is the best estimate for 4.7×8?

(A) 32

(B) 40

(C) 48

(D) 36

9. What is $5.00 \div 0.25$?

(A) 2

(B) 25

(C) 20

(D) 200

10. A recipe calls for $\frac{2}{3}$ cup of milk and $\frac{1}{4}$ cup of cream. How much liquid is needed in all?

(A) $\frac{3}{7}$ cup

(B) $\frac{11}{12}$ cup

(C) $\frac{5}{12}$ cup

(D) $\frac{7}{12}$ cup

11. Which benchmark is $\frac{7}{8}$ closest to?

(A) 0

(B) $\frac{1}{2}$

(C) 1

(D) 2

Find more at
ViewMath.com/ND-Grade5

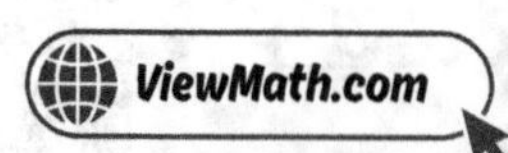

12. Three friends shared a long sub sandwich. Ana ate $\frac{1}{3}$, Ben ate $\frac{1}{4}$, and Cara ate $\frac{1}{6}$. What fraction of the sandwich is left?

(A) $\frac{1}{4}$

(B) $\frac{3}{4}$

(C) $\frac{1}{3}$

(D) $\frac{1}{12}$

13. Four friends share 3 pizzas equally. How much pizza does each friend get?

(A) $\frac{4}{3}$ of a pizza

(B) $1\frac{1}{3}$ pizzas

(C) $\frac{3}{4}$ of a pizza

(D) $\frac{1}{4}$ of a pizza

14. A bag weighs $\frac{3}{4}$ pound. What is the weight of 5 bags?

Your Answer:

15. What is $\frac{2}{3} \times \frac{3}{2}$?

(A) $\frac{5}{5}$

(B) $\frac{4}{9}$

(C) 1

(D) $\frac{6}{5}$

16. Jake runs 5 miles. Tomorrow he plans to run $\frac{6}{5}$ times as far. Without computing, will he run more or less than 5 miles?

(A) Less, because $\frac{6}{5}$ has a 5 in it.

(B) More, because $\frac{6}{5} > 1$.

(C) The same distance.

(D) Less, because 6 and 5 are close.

Find more at
ViewMath.com/ND-Grade5

17. *What is* $4 \times 3\frac{1}{2}$?

(A) 12

(B) $12\frac{1}{2}$

(C) 14

(D) $7\frac{1}{2}$

18. *A parking lot has 48 spaces.* $\frac{5}{8}$ *of the spaces are full. How many cars are parked?*

(A) 24

(B) 30

(C) 36

(D) 40

19. *What is* $\frac{1}{3} \div 4$?

(A) $\frac{4}{3}$

(B) $\frac{1}{7}$

(C) $\frac{1}{12}$

(D) $\frac{3}{4}$

20. *Evaluate:* $48 \div (2 + 6)$

(A) 30

(B) 8

(C) 6

(D) 5

21. *A bakery charges $3 per cupcake plus a $5 box fee. What is the total cost for 8 cupcakes in a box?*

Your Answer:

22. *Corresponding terms from two patterns are* $A = 8$ *and* $B = 24$. *What is the relationship?*

(A) *B is 2 times A*

(B) *B is 3 times A*

(C) *B is 4 times A*

(D) *B is A plus 16*

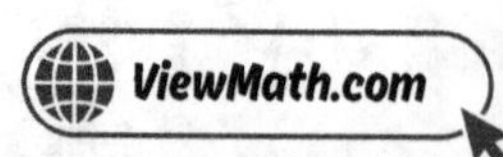

23. Carlos starts at the origin. He walks 7 units to the right and 5 units up. What ordered pair describes his location?

Your Answer:

24. A dog eats 3 cups of food per day. Write the first 5 ordered pairs (days, cups) starting at day 0.

Your Answer:

25. Convert 84 inches to feet.

Your Answer:

26. A line plot shows the following data for the lengths of worms (in inches):

$\frac{1}{4}$: 3 X marks $\frac{1}{2}$: 5 X marks $\frac{3}{4}$: 2 X marks   1: 1 X mark

How many worms were measured in all?

(A) 4   (B) 9

(C) 11   (D) 10

27. A unit cube has edges that are each 1 cm long. What is its volume?

(A) 1 cm   (B) 1 cm^2

(C) 1 cm^3 (D) 3 cm

28. A rectangular block is $12 \times 6 \times 4$ in. A $4 \times 2 \times 4$ in notch is removed. What is the volume?

(A) 256 in^3 (B) 288 in^3

(C) 320 in^3 (D) 224 in^3

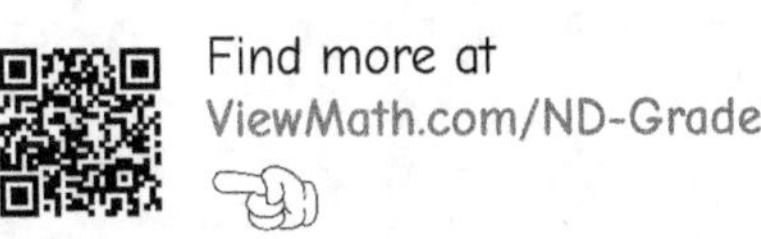

Find more at
ViewMath.com/ND-Grade5

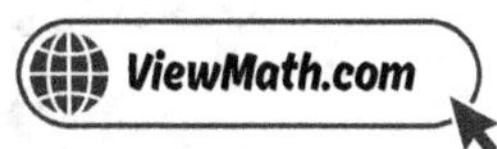

29. Which is the best estimate for the volume of a box that is $11 \times 9 \times 4$ inches?

(A) $200\ in^3$

(B) $400\ in^3$

(C) $600\ in^3$

(D) $1{,}000\ in^3$

30. Is the statement "All parallelograms are rectangles" always true, sometimes true, or never true?

Your Answer:

★ *End of Practice Test 10* ★

Great job finishing the test!

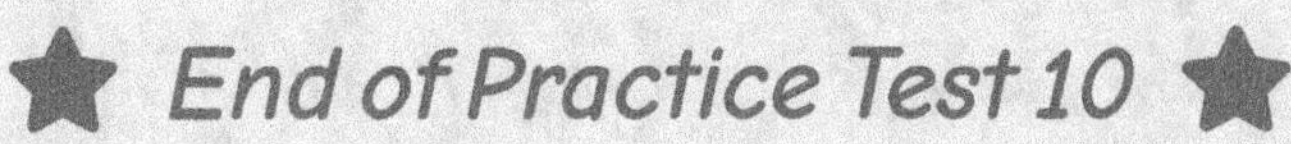

✅ My Score

I got ___________ out of 30 questions right.

*Check your answers in the **Answer Key** at the back of the book.*

💡 *Review any questions you missed. That's how we learn!*

📊 Check Your Score Online!

Visit **ViewMath Academy** to enter your answers and see which topics you need to review. You can also explore lessons, take quizzes, track your scores, and save your progress!

viewmath.com/score/5.1.ND.25

Or go to viewmath.com/score and enter code: **5.1.ND.25**

Answer Key & Explanations

⭐ Check Your Answers! ⭐

First try each test on your own, then look here to check.

Read the explanations to learn from any mistakes ⭐

✅ Practice Test 1 — Answer Key

1 C	2 C	3 7.3	4 B	5 A	6 80	7 C	8 C	9 7	10 $\frac{7}{9}$
11 B	12 D	13 B	14 D	15 C	16 B	17 3	18 C	19 C	20 C

21 Divide 24 by 8, then add 5.	22 $(0,0), (3,12), (6,24), (9,36)$	23 B	24 C	25 5 yd

26 B	27 D	28 620 cm^3	29 180 m^3	30 C

💡 Time to Learn! 💡

Go through the explanations below, **especially for the questions you missed**.

Understanding why each answer is correct makes you a stronger math thinker!

👍 **Tip:** Circle any questions you got wrong, then read their explanation carefully.

📖 Practice Test 1 — Detailed Explanations

1. $10^3 = 10 \times 10 \times 10 = 1,000$. The exponent 3 tells you there are 3 zeros after the 1.

Find more at
ViewMath.com/ND-Grade5

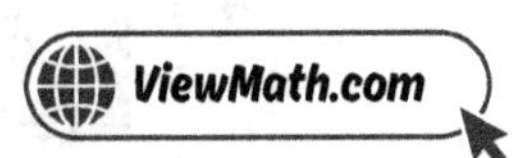

2. The last digit of 0.47 is in the hundredths place, so the denominator is 100. Thus $0.47 = \frac{47}{100}$.

3. The tenths digit is 3. Look at only the hundredths digit (4, the digit immediately to the right). Since $4 < 5$, round down: keep tenths as 3. Answer: 7.3.

4. The area model breaks both numbers into expanded form: $(50 + 2) \times (40 + 7)$, which gives four partial products.

5. $425 \div 30 = 14$ R5. There will be 14 full boxes.

6. $4{,}780 \approx 4{,}800$ and $58 \approx 60$. So $4{,}800 \div 60 = 80$.

7. Annex a zero: $8.30 - 5.17$. Hundredths: $0 < 7$, borrow, $10 - 7 = 3$. Tenths: $2 - 1 = 1$. Ones: $8 - 5 = 3$. Answer: 3.13.

8. $0.75 \times 8 = 6.00$.

9. Multiply both by 10: $63 \div 9 = 7$.

10. LCD is 9. $\frac{1}{3} = \frac{3}{9}$. So $\frac{4}{9} + \frac{3}{9} = \frac{7}{9}$.

11. $\frac{2}{5} \approx \frac{1}{2}$ and $\frac{1}{2} = \frac{1}{2}$. Estimate: $\frac{1}{2} + \frac{1}{2} = 1$. Since $\frac{3}{7} < \frac{1}{2}$, the answer is not reasonable.

12. "Remaining" means how much is left, which requires subtraction. "Total," "combined," and "altogether" are clue words for addition.

13. The fraction $\frac{a}{b} = a \div b$. So $\frac{5}{8} = 5 \div 8$.

14. $\frac{1}{4} \times 12 = \frac{12}{4} = 3$, which is a whole number. The others do not simplify to whole numbers.

Find more at
ViewMath.com/ND-Grade5

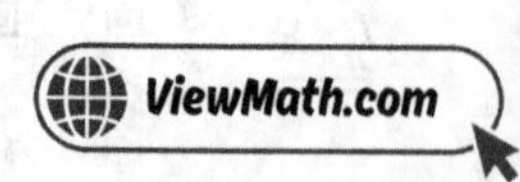

15 Area $= \frac{3}{4} \times \frac{2}{5} = \frac{6}{20} = \frac{3}{10}$ square inch.

16 Multiplying by $\frac{1}{2}$ is the same as dividing by 2. Both give half the number.

17 $\frac{5}{2} \times \frac{6}{5} = \frac{30}{10} = 3$.

18 $\frac{1}{3} \times 60 = 20$ sold. Remaining: $60 - 20 = 40$.

19 $\frac{1}{4} \div 5 = \frac{1}{4 \times 5} = \frac{1}{20}$.

20 Innermost first: $4 \times 3 = 12$. Brackets: $20 - 12 = 8$. Then $8 + 7 = 15$.

21 The parentheses show dividing first, then adding 5.

22 A: $0, 3, 6, 9$. B: $0, 12, 24, 36$. Pair up corresponding terms.

23 The x-coordinate tells how far right. $(9, 1)$ has the largest x-coordinate (9).

24 Width: $6 - 1 = 5$. Height: $4 - 1 = 3$. Since the sides are not equal, it is a rectangle.

25 $15 \div 3 = 5$ yd.

26 A line plot shows how many data points fall at each value, making it easy to see frequency.

27 Bottom layer: $6 \times 2 = 12$. Total: $12 \times 4 = 48$ cubic units.

28 Full: $14 \times 10 \times 5 = 700$. Removed: $4 \times 4 \times 5 = 80$. Remaining: $700 - 80 = 620 \ cm^3$.

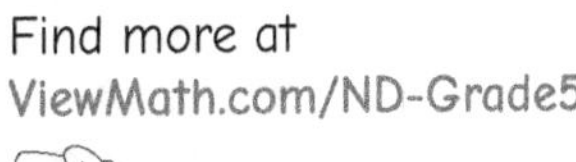

Find more at
ViewMath.com/ND-Grade5

29   $V = 15 \times 6 \times 2 = 180 \ m^3.$

30   4 equal sides and 2 pairs of parallel sides make it a parallelogram and a rhombus. No right angles means it's not a rectangle or square.

Practice Test 2 — Answer Key

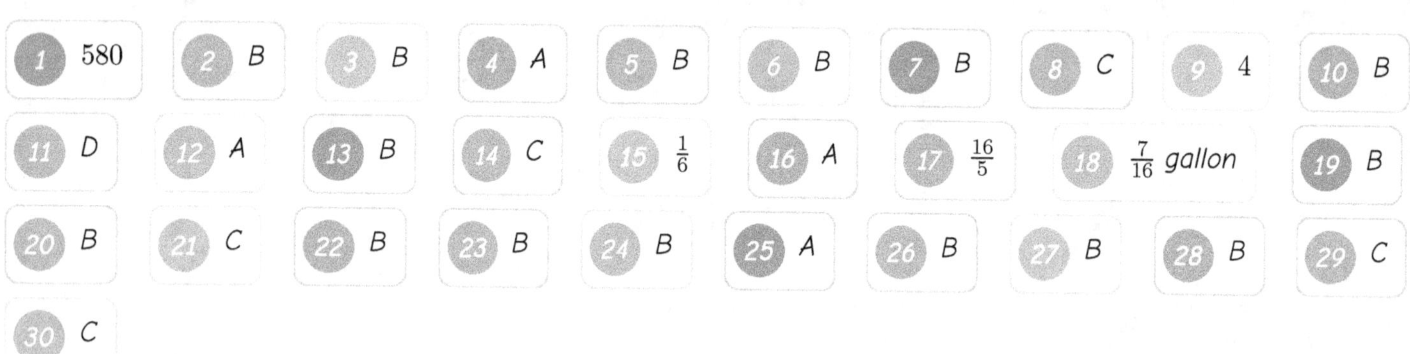

1 580	2 B	3 B	4 A	5 B	6 B	7 B	8 C	9 4	10 B
11 D	12 A	13 B	14 C	15 $\frac{1}{6}$	16 A	17 $\frac{16}{5}$	18 $\frac{7}{16}$ gallon	19 B	
20 B	21 C	22 B	23 B	24 B	25 A	26 B	27 B	28 B	29 C
30 C									

💡 Time to Learn! 💡

Go through the explanations below, **especially for the questions you missed**.

Understanding why each answer is correct makes you a stronger math thinker!

👍 *Tip:* Circle any questions you got wrong, then read their explanation carefully.

📖 Practice Test 2 — Detailed Explanations

1   $5.8 \times 100 = 580.$ The decimal point moves 2 places to the right: $5.8 \rightarrow 580.$

2   The digit 7 is in the tenths place (first place after the decimal). The word form is "seven tenths."

Find more at
ViewMath.com/ND-Grade5

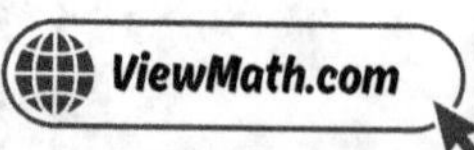

3 The ones digit is 0. Look at the tenths digit (9). Since $9 \geq 5$, round up the ones digit: $0 + 1 = 1$. Answer: 1.

4 $56 \times 3 = 168$ and $56 \times 20 = 1{,}120$. $168 + 1{,}120 = 1{,}288$.

5 $85 \div 6 = 14$ R1. The question asks for the leftover amount, which is the remainder, 1.

6 $78 \approx 80$ and $51 \approx 50$. So $80 \times 50 = 4{,}000$.

7 Write 10 as 10.00. Then $10.00 - 3.47 = 6.53$.

8 $0.25 \times 6 = 1.50$ miles.

9 Multiply both by 10: $32 \div 8 = 4$.

10 LCD of 4 and 3 is 12. $\frac{3}{4} = \frac{9}{12}$ and $\frac{1}{3} = \frac{4}{12}$. So $\frac{9}{12} + \frac{4}{12} = \frac{13}{12} = 1\frac{1}{12}$. Choices B and D are equivalent.

11 $\frac{7}{8} \approx 1$ and $\frac{4}{5} \approx 1$. Estimate: $1 + 1 = 2$.

12 Added: $4\frac{5}{8} + 3\frac{1}{2} = 4\frac{5}{8} + 3\frac{4}{8} = 7\frac{9}{8} = 8\frac{1}{8}$. Needed: $12 - 8\frac{1}{8} = 11\frac{8}{8} - 8\frac{1}{8} = 3\frac{7}{8}$.

13 $8 \div 5 = \frac{8}{5} = 1\frac{3}{5}$ apples each.

14 $\frac{3 \times 8}{4} = \frac{24}{4} = 6$.

15 $\frac{1}{5} \times \frac{5}{6} = \frac{5}{30} = \frac{1}{6}$.

16 $\frac{11}{10} > 1$ (since $11 > 10$), so the product is greater than 14.

Find more at
ViewMath.com/ND-Grade5

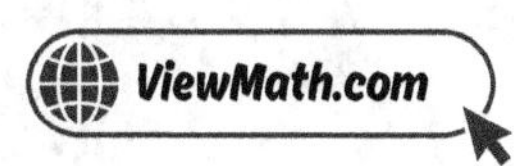

17 $3\frac{1}{5} = \frac{3\times5+1}{5} = \frac{16}{5}$.

18 $\frac{1}{2} \times \frac{7}{8} = \frac{7}{16}$ gallon.

19 If $\frac{1}{3} \div 2 = \frac{1}{6}$, then $\frac{1}{6} \times 2$ should equal $\frac{1}{3}$. $\frac{1}{6} \times 2 = \frac{2}{6} = \frac{1}{3}$. ✓

20 Parentheses: $1 + 2 = 3$. Brackets: $3 \times 3 = 9$. Braces: $5 + 9 = 14$. Then $2 \times 14 = 28$.

21 "The sum of 10 and 5" is $(10 + 5)$. "Twice" means $\times\, 2$: $(10 + 5) \times 2 = 30$.

22 A: $0, 4, 8$. B: $0, 12, 24$. The 3rd pair is $(8, 24)$.

23 8 right = x-coordinate 8, 3 up = y-coordinate 3. The pair is $(8, 3)$.

24 $5 per week: week $1 = \$5$, week $2 = \$10$, week $3 = \$15$.

25 $10{,}000 \div 2{,}000 = 5$ T.

26 $2 \times \frac{1}{4} = \frac{1}{2}$; $3 \times \frac{3}{8} = \frac{9}{8}$; $3 \times \frac{1}{2} = \frac{3}{2}$. Total: $\frac{4}{8} + \frac{9}{8} + \frac{12}{8} = \frac{25}{8} = 3\frac{1}{8}$.

27 A $1 \times 1 \times 12$ box and a $2 \times 2 \times 3$ box both hold 12 cubes but have different shapes.

28 Main: $12 \times 6 \times 2 = 144$. Smaller: $4 \times 3 \times 2 = 24$. Total: $144 + 24 = 168$ m^3.

29 Box: $10 \times 8 \times 6 = 480$ in^3. Small cube: $2 \times 2 \times 2 = 8$ in^3. $480 \div 8 = 60$ cubes.

30 A quadrilateral with exactly one pair of parallel sides is a trapezoid (but not a parallelogram).

Find more at
ViewMath.com/ND-Grade5

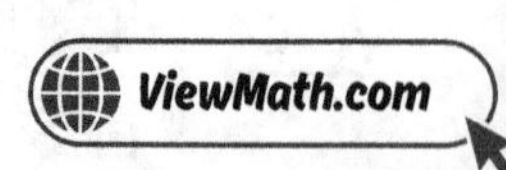

☑ Practice Test 3 — Answer Key

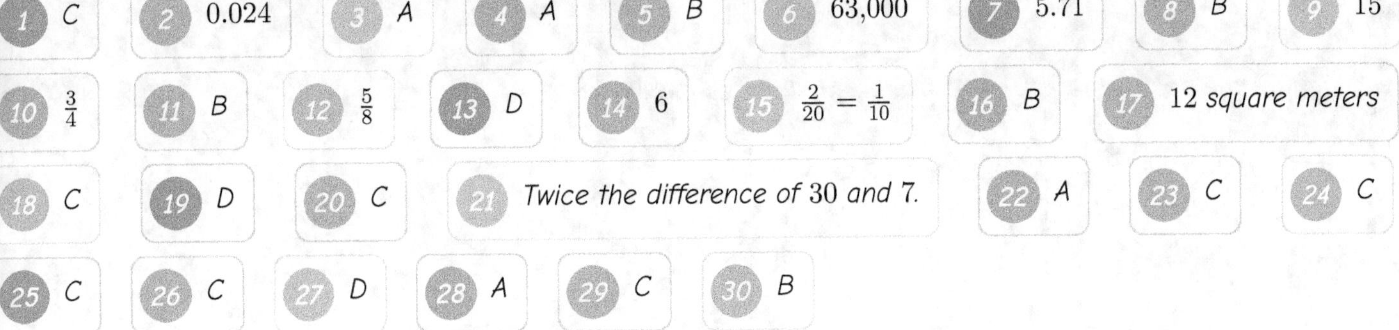

1 C	**2** 0.024	**3** A
4 A	**5** B	**6** 63,000
7 5.71	**8** B	**9** 15
10 $\frac{3}{4}$	**11** B	**12** $\frac{5}{8}$
13 D	**14** 6	**15** $\frac{2}{20} = \frac{1}{10}$
16 B	**17** 12 square meters	
18 C	**19** D	**20** C
21 Twice the difference of 30 and 7.	**22** A	**23** C
24 C	**25** C	**26** C
27 D	**28** A	**29** C
30 B		

💡 Time to Learn! 💡

Go through the explanations below, **especially for the questions you missed**.

Understanding why each answer is correct makes you a stronger math thinker!

👍 **Tip:** Circle any questions you got wrong, then read their explanation carefully.

📖 Practice Test 3 — Detailed Explanations

1 Dividing by 100 moves the decimal point 2 places to the left: $700 \rightarrow 7$.

2 Thousandths means the denominator is 1,000. $\frac{24}{1,000} = 0.024$ (the last digit 4 is in the thousandths place).

3 The tenths digit is 3. Look at the hundredths digit (7). Since $7 \geq 5$, round up: $3 + 1 = 4$. Answer: 6.4.

4 $5,009 \times 2 = 10,018$ and $5,009 \times 10 = 50,090$. $10,018 + 50,090 = 60,108$.

5 $145 \div 8 = 18$ R1. Since 1 student still needs a ride, you must round up to 19 vans.

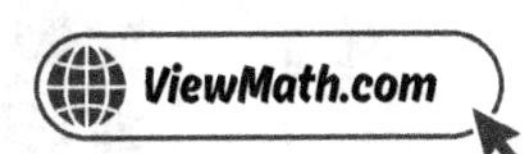

6. $715 \approx 700$ and $89 \approx 90$. So $700 \times 90 = 63{,}000$.

7. $10.00 - 4.29 = 5.71$.

8. *Ignore the decimal:* $4 \times 9 = 36$. *Two decimal places in* 0.04, *so* $36 \to 0.36$.

9. *Multiply both by* 100: $750 \div 50 = 15$.

10. *LCD of* 10, 4, *and* 5 *is* 20. $\frac{3}{10} = \frac{6}{20}$, $\frac{1}{4} = \frac{5}{20}$, $\frac{1}{5} = \frac{4}{20}$. *So* $\frac{6}{20} + \frac{5}{20} + \frac{4}{20} = \frac{15}{20} = \frac{3}{4}$.

11. $\frac{7}{8} \approx 1$, *so* $3\frac{7}{8} \approx 4$. $\frac{1}{5} \approx 0$, *so* $2\frac{1}{5} \approx 2$. *Estimate:* $4 + 2 = 6$.

12. *LCD is* 8. $\frac{1}{4} = \frac{2}{8}$. $\frac{3}{8} + \frac{2}{8} = \frac{5}{8}$.

13. *4 packs shared among 7 kids is* $4 \div 7 = \frac{4}{7}$.

14. $\frac{2 \times 9}{3} = \frac{18}{3} = 6$.

15. $\frac{1 \times 2}{4 \times 5} = \frac{2}{20} = \frac{1}{10}$.

16. $\frac{3}{4} < 1$, *so multiplying by* $\frac{3}{4}$ *makes the result smaller than* 20.

17. $\frac{9}{2} \times \frac{8}{3} = \frac{72}{6} = 12$ *square meters*.

18. $\frac{3}{8} \times 40 = \frac{120}{8} = 15$ *goldfish*.

19. $\frac{1}{6} \div 2 = \frac{1}{6 \times 2} = \frac{1}{12}$.

Find more at
ViewMath.com/ND-Grade5

20 Parentheses first: $12 - 7 = 5$. Then $5 \times 5 = 25$.

21 $(30 - 7)$ is the difference. Multiplying by 2 gives "twice the difference."

22 $5 - 2 = 3$, $8 - 5 = 3$, $11 - 8 = 3$. The rule is start at 2, add 3.

23 3 units right means $x = 3$. 0 units up means $y = 0$. So $R = (3, 0)$.

24 A: $0, 1, 2, 3$. B: $0, 5, 10, 15$. When $A = 3$ (the 4th term), $B = 15$.

25 1 cup $= 8$ fl oz, so $4 \times 8 = 32$ fl oz.

26 Range $= \frac{7}{8} - \frac{1}{8} = \frac{6}{8} = \frac{3}{4}$ inch.

27 Bottom layer: $4 \times 3 = 12$ cubes. Total: $12 \times 5 = 60$ cubic units.

28 Full: $8 \times 6 \times 5 = 240$. Removed: $4 \times 2 \times 5 = 40$. U-shape: $240 - 40 = 200$.

29 $h = V \div B = 252 \div 42 = 6$ in.

30 A slanted parallelogram does not have right angles. Only rectangles (and squares, which are special rectangles) have 4 right angles.

Practice Test 4 — Answer Key

 **1** B    **2** 24.007    **3** B    **4** 48,240    **5** B    **6** B    **7** A    **8** 5.85   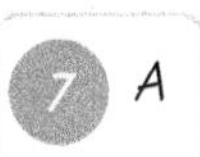 **9** B

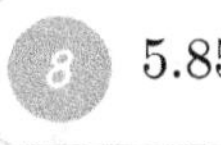

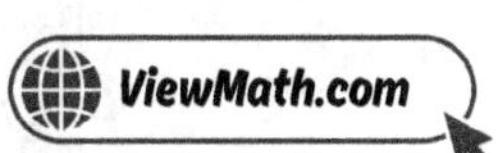

10 $\frac{7}{8}$ gallon	**11** 4	**12** $1\frac{5}{12}$ meters	**13** $\frac{2}{3}$	**14** D	**15** $\frac{1}{2}$	**16** C	**17** B
18 3 square inches	**19** D	**20** B	**21** B	**22** B	**23** B	**24** 6	**25** C	**26** $4\frac{3}{4}$
27 B	**28** B	**29** C	**30** Rectangle

💡 **Time to Learn!** 💡

*Go through the explanations below, **especially for the questions you missed**.*

Understanding why each answer is correct makes you a stronger math thinker!

👍 *Tip: Circle any questions you got wrong, then read their explanation carefully.*

📖 Practice Test 4 — Detailed Explanations

1 Multiplying by $10^2 = 100$ shifts every digit 2 places to the left on the place value chart, which is the same as moving the decimal point 2 places to the right.

2 $2 \times 10 = 20$, $4 \times 1 = 4$, whole part $= 24$. $7 \times \frac{1}{1,000} = 0.007$. Standard form: 24.007.

3 The hundredths digit is 5. Since $5 \geq 5$, round up the tenths digit: $3 + 1 = 4$. The correct answer is 4.4.

4 $3,015 \times 6 = 18,090$ and $3,015 \times 10 = 30,150$. $18,090 + 30,150 = 48,240$.

5 $300 \div 45 = 6$ R30. The truck must make 7 trips to deliver all the boxes.

6 $3,150 \approx 3,200$ and $42 \approx 40$. So $3,200 \div 40 = 80$.

7 $8.50 - 3.75 = 4.75$ meters.

Find more at
ViewMath.com/ND-Grade5

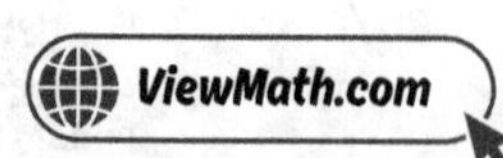

8 $65 \times 9 = 585$. *Two decimal places:* $585 \to 5.85$.

9 *Multiply both by* 100: $120 \div 4 = 30$.

10 *LCD is* 8. $\frac{1}{2} = \frac{4}{8}$. *So* $\frac{3}{8} + \frac{4}{8} = \frac{7}{8}$.

11 $\frac{7}{8} \approx 1$, *so* $5\frac{7}{8} \approx 6$. $\frac{1}{6} \approx 0$, *so* $2\frac{1}{6} \approx 2$. *Estimate:* $6 - 2 = 4$.

12 *LCD of 4 and 6 is* 12. $\frac{1}{4} = \frac{3}{12}$ *and* $\frac{5}{6} = \frac{10}{12}$. *Borrow:* $3\frac{3}{12} = 2\frac{15}{12}$. $\frac{15}{12} - \frac{10}{12} = \frac{5}{12}$. *Wholes:* $2 - 1 = 1$. *Answer:* $1\frac{5}{12}$.

13 $2 \div 3 = \frac{2}{3}$ *of a pie.*

14 $\frac{1 \times 7}{2} = \frac{7}{2} = 3\frac{1}{2}$.

15 $\frac{4 \times 5}{5 \times 8} = \frac{20}{40} = \frac{1}{2}$.

16 $\frac{6}{6} = 1$. *Multiplying by 1 does not change the value.*

17 $\frac{7}{4} \times \frac{4}{7} = \frac{28}{28} = 1$.

18 $\frac{9}{4} \times \frac{4}{3} = \frac{36}{12} = 3$ *square inches.*

19 $\frac{1}{6} \div 3 = \frac{1}{18}$ *acre.*

20 $(6 + 2) \times 4 = 8 \times 4 = 32$.

21 *"Add 12 and 8" first:* $(12 + 8)$. *"Then subtract 5":* $(12 + 8) - 5$.

Find more at
ViewMath.com/ND-Grade5

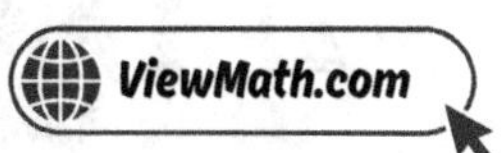

22 $12 \div 4 = 3$. Leo earns 3 times more than Ava.

23 A point is on the y-axis when its x-coordinate is 0. Only $(0, 6)$ has $x = 0$.

24 From $(2, 6)$ to $(4, 12)$: y increases by 6. From $(4, 12)$ to $(6, 18)$: y increases by 6.

25 $1\ pt = 2$ cups, so $3 \times 2 = 6$ cups.

26 $5 \times \frac{1}{2} = \frac{5}{2} = 2\frac{1}{2}$. $3 \times \frac{3}{4} = \frac{9}{4} = 2\frac{1}{4}$. Total: $2\frac{1}{2} + 2\frac{1}{4} = 4\frac{3}{4}$.

27 $12 \times 3 = 36$ cubic units.

28 Each prism: $5 \times 3 \times 4 = 60$. Total: $60 + 60 = 120$ cubic units.

29 $V = 4 \times 3 \times 2 = 24\ ft^3$.

30 It has 4 right angles, so it's a rectangle. But the sides are not all equal (8 cm and 5 cm), so it is not a square.

✓ Practice Test 5 — Answer Key

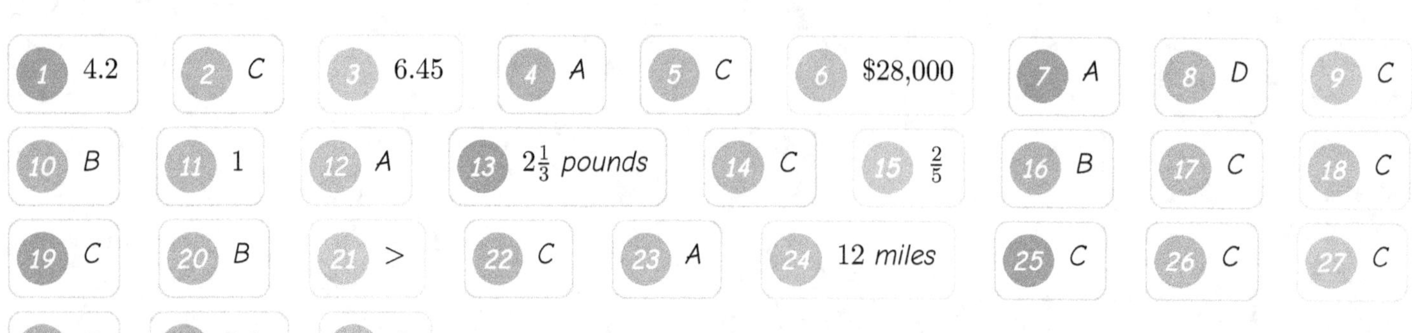

1	4.2	2	C	3	6.45	4	A	5	C	6	$28,000	7	A	8	D	9	C
10	B	11	1	12	A	13	$2\frac{1}{3}$ pounds	14	C	15	$\frac{2}{5}$	16	B	17	C	18	C
19	C	20	B	21	>	22	C	23	A	24	12 miles	25	C	26	C	27	C
28	B	29	6 in	30	D

Find more at
ViewMath.com/ND-Grade5

ViewMath.com

💡 Time to Learn! 💡

Go through the explanations below, **especially for the questions you missed**.

Understanding why each answer is correct makes you a stronger math thinker!

👍 *Tip:* Circle any questions you got wrong, then read their explanation carefully.

📖 Practice Test 5 — Detailed Explanations

1. $4{,}200 \div 1{,}000 = 4.2$. The decimal point moves 3 places to the left: $4{,}200 \to 4.2$.

2. A denominator of 1,000 means thousandths. $\frac{7}{1{,}000} = 0.007$ (the 7 is in the thousandths place).

3. To round to 6.5, the hundredths digit must be 5 or more. The smallest two-decimal number with that property is 6.45 (hundredths = 5, rounds up tenths: $4 + 1 = 5$).

4. $2{,}841 \times 1 = 2{,}841$ and $2{,}841 \times 30 = 85{,}230$. $2{,}841 + 85{,}230 = 88{,}071$.

5. $15 \div 4 = 3 \text{ R}3$. Since brownies can be split, the remainder is shared: $3\frac{3}{4}$ brownies each.

6. $68 \approx 70$ and $415 \approx 400$. So $70 \times 400 = 28{,}000$.

7. Hundredths: $7 - 3 = 4$. Tenths: $8 - 2 = 6$. Ones: $9 - 4 = 5$. Answer: 5.64.

8. Ignore the decimal: $408 \times 7 = 2{,}856$. Two decimal places in 4.08, so $2856 \to 28.56$.

9. Multiply both by 10: $75.6 \div 9 = 8.4$.

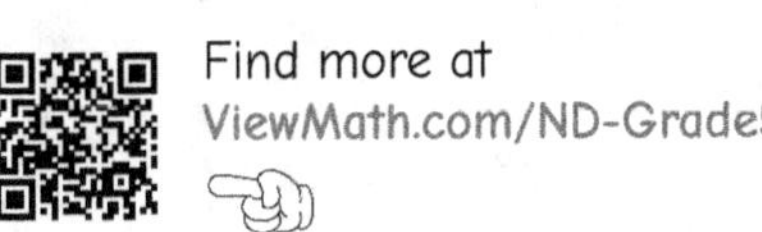
Find more at
ViewMath.com/ND-Grade5

10 LCD is 6. $\frac{2}{3} = \frac{4}{6}$. So $\frac{1}{6} + \frac{4}{6} = \frac{5}{6}$.

11 The numerator 9 is very close to the denominator 10, so $\frac{9}{10}$ rounds to 1.

12 Total used: $1\frac{1}{4} + 2\frac{1}{3}$. LCD of 4 and 3 is 12. $\frac{1}{4} = \frac{3}{12}$ and $\frac{1}{3} = \frac{4}{12}$. Used $= 3\frac{7}{12}$. Left: $5 - 3\frac{7}{12} = 4\frac{12}{12} - 3\frac{7}{12} = 1\frac{5}{12}$.

13 $7 \div 3 = \frac{7}{3} = 2\frac{1}{3}$ pounds.

14 $\frac{2}{3} \times 12 = \frac{24}{3} = 8$ meters.

15 $\frac{3 \times 2}{5 \times 3} = \frac{6}{15} = \frac{2}{5}$.

16 $\frac{2}{3} < 1$, so multiplying by it makes the price lower.

17 $\frac{3}{2} \times \frac{7}{3} = \frac{21}{6} = \frac{7}{2} = 3\frac{1}{2}$.

18 $\frac{1}{3} \times 24 = 8$ cookies.

19 Dividing by 5 is the same as multiplying by $\frac{1}{5}$. So $\frac{1}{2} \div 5 = \frac{1}{2} \times \frac{1}{5} = \frac{1}{10}$.

20 B: $(7 + 3) \times 2 = 20$, but $7 + 3 \times 2 = 13$. The other pairs all give the same answer.

21 $46 + 12$ is greater than 46, so 8 times a bigger number is greater.

22 Each term increases by 5: $0 + 5 = 5$, $5 + 5 = 10$, and so on.

23 $(1, 1)$ is only 1 unit right and 1 unit up, making it nearest to $(0, 0)$.

Find more at
ViewMath.com/ND-Grade5

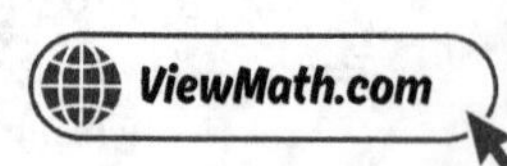

24 A: $4 \times 2 = 8$ miles. B: $4 \times 5 = 20$ miles. Difference: $20 - 8 = 12$ miles.

25 1 ft = 12 in, so $5 \times 12 = 60$ in.

26 $6 \times \frac{1}{4} = \frac{6}{4} = 1\frac{1}{2}$. $1 \times \frac{3}{4} = \frac{3}{4}$. Total: $1\frac{1}{2} + \frac{3}{4} = 1\frac{2}{4} + \frac{3}{4} = 2\frac{1}{4}$.

27 Volume measures the amount of space inside a three-dimensional shape.

28 Large: $9 \times 5 \times 4 = 180$. Small: $3 \times 5 \times 2 = 30$. Total: $180 + 30 = 210$.

29 $6 \times 6 \times 6 = 216$, so each edge is 6 inches.

30 Rectangles have 4 right angles, opposite sides equal, and 2 pairs of parallel sides. But not all rectangles have 4 equal sides — only squares do.

Practice Test 6 — Answer Key

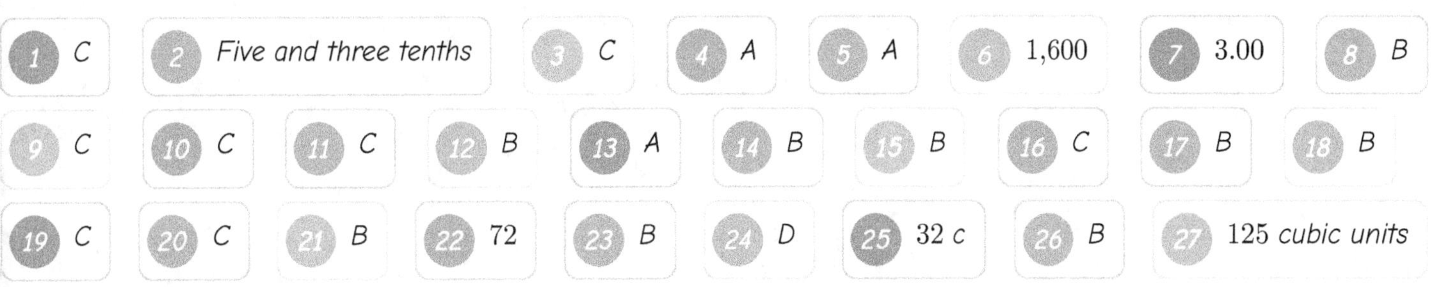

1	C	2	Five and three tenths	3	C	4	A	5	A	6	1,600	7	3.00	8	B
9	C	10	C	11	C	12	B	13	A	14	B	15	B	16	C	17	B	18	B
19	C	20	C	21	B	22	72	23	B	24	D	25	32 c	26	B	27	125 cubic units
28	C	29	720 in^3	30	C

Find more at
ViewMath.com/ND-Grade5

> 💡 *Time to Learn!* 💡
>
> *Go through the explanations below, **especially for the questions you missed.***
>
> *Understanding why each answer is correct makes you a stronger math thinker!*
>
> 👍 *Tip: Circle any questions you got wrong, then read their explanation carefully.*

📖 Practice Test 6 — Detailed Explanations

1. $0.7 \times 10^3 = 0.7 \times 1,000 = 700$. *The decimal point moves 3 places to the right:* $0.7 \to 700$.

2. *The whole part is 5 ("five"), "and" marks the decimal, the 3 is in the tenths place: "five and three tenths."*

3. *The tenths digit is 9. Look at the hundredths digit (6). Since* $6 \geq 5$, *round up:* $9 + 1 = 10$ *tenths* $= 1$ *whole. Carry to the ones:* $9 + 1 = 10.0$.

4. $73 \times 9 = 657$ *and* $73 \times 90 = 6,570$. $657 + 6,570 = 7,227$.

5. $75 \div 22 = 3$ R9. *Stickers cannot be split, so each student gets 3 stickers.*

6. $83 \approx 80$ *and* $21 \approx 20$. *So* $80 \times 20 = 1,600$.

7. $8.45 - 5.45 = 3.00$.

8. *Ignore the decimal:* $6 \times 4 = 24$. *Two decimal places in 0.06, so* $24 \to 0.24$.

9. *Multiply both by 10:* $36 \div 4 = 9$.

Find more at
ViewMath.com/ND-Grade5

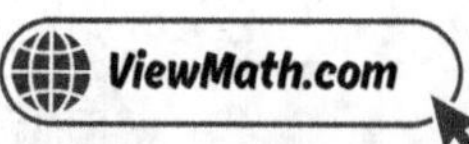

10 LCD is 10. $\frac{2}{5} = \frac{4}{10}$. So $\frac{3}{10} + \frac{4}{10} = \frac{7}{10}$.

11 $\frac{3}{4} + \frac{7}{8} \approx 1 + 1 = 2$. $\frac{2}{5} + \frac{1}{3} \approx \frac{1}{2} + \frac{1}{2} = 1$. The first sum is greater.

12 "How much farther" means subtract. LCD of 4 and 3 is 12. $\frac{3}{4} = \frac{9}{12}$ and $\frac{2}{3} = \frac{8}{12}$. So $\frac{9}{12} - \frac{8}{12} = \frac{1}{12}$.

13 $9 \div 4 = \frac{9}{4} = 2\frac{1}{4}$ feet.

14 Ben should only multiply the numerator. $\frac{2 \times 5}{7} = \frac{10}{7}$. He incorrectly multiplied the denominator too.

15 $\frac{5 \times 3}{6 \times 5} = \frac{15}{30} = \frac{1}{2}$.

16 $\frac{6}{6} = 1$, so the product equals 45.

17 $\frac{5}{4} \times \frac{4}{3} = \frac{20}{12} = \frac{5}{3} = 1\frac{2}{3}$.

18 $\frac{2}{3} \times \frac{9}{2} = \frac{18}{6} = 3$ feet.

19 $\frac{1}{4} \div 3 = \frac{1}{12}$ of a pizza.

20 Left: $(8 + 2) \times 3 = 10 \times 3 = 30$. Right: $8 + 2 \times 3 = 8 + 6 = 14$.

21 "Add 8 and 7" goes in parentheses: $(8 + 7)$. "Then multiply by 2" gives $2 \times (8 + 7)$.

22 B: $0, 18, 36, 54, 72$. The 5th term is 72.

23 The x-coordinate comes first: $(0, 8)$.

Find more at
ViewMath.com/ND-Grade5

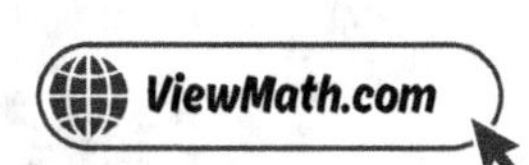

24 Amy: $4 \times 10 = 40$ pages. Ben: $4 \times 5 = 20$ pages. Difference: $40 - 20 = 20$.

25 1 gal $= 16$ cups, so $2 \times 16 = 32$ cups.

26 Greatest: $\frac{3}{8}$. Least: $\frac{1}{8}$. Difference: $\frac{3}{8} - \frac{1}{8} = \frac{2}{8} = \frac{1}{4}$.

27 $5 \times 5 \times 5 = 125$ cubic units.

28 $V_A = 7 \times 3 \times 5 = 105$. $V_B = 4 \times 3 \times 5 = 60$. Total: $105 + 60 = 165$.

29 $\frac{3}{4}$ of $10 = 7.5$ in. $V = 12 \times 8 \times 7.5 = 720$ in^3.

30 By definition, a rhombus has all 4 sides equal. The other statements are only sometimes true.

Practice Test 7 — Answer Key

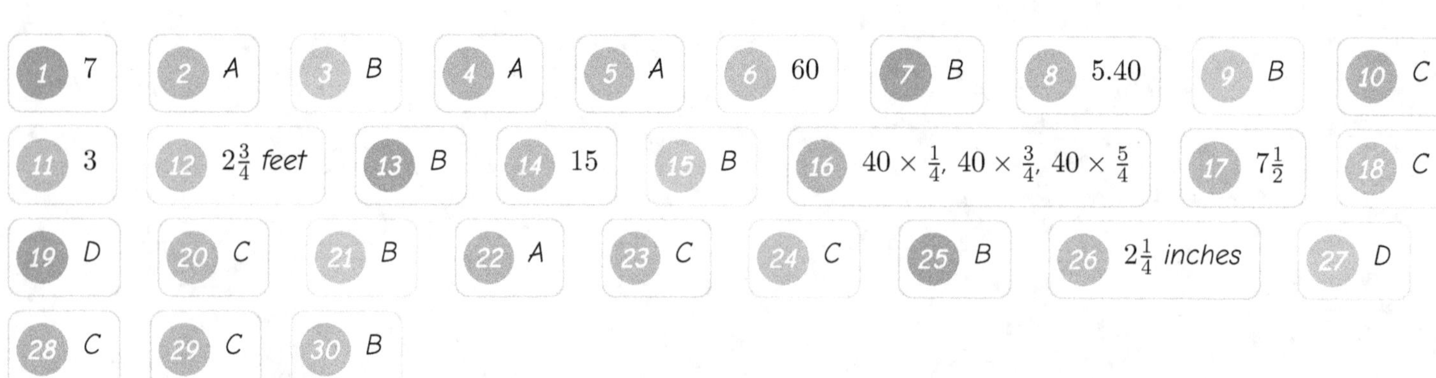

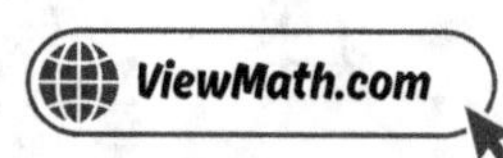

💡 Time to Learn! 💡

Go through the explanations below, **especially for the questions you missed.**

Understanding why each answer is correct makes you a stronger math thinker!

👍 *Tip:* Circle any questions you got wrong, then read their explanation carefully.

📖 Practice Test 7 — Detailed Explanations

1 $7 \times 1{,}000 = 7{,}000$. *To undo* $\times 10^3$, *divide:* $7{,}000 \div 1{,}000 = 7$.

2 $4 \times 10 = 40$ *and* $8 \times 1 = 8$, *so the whole part is* 48. $3 \times \frac{1}{100} = 0.03$. *Standard form:* 48.03.

3   *The tenths digit is* 9. *Look at only the hundredths digit* (9). *Since* $9 \geq 5$, *round up:* $9 + 1 = 10$ *tenths. Carry to the ones:* $9 + 1 = 10.0$.

4 $350 \times 24 = 8{,}400$.

5 $50 \div 6 = 8$ *R2. She can only buy full notebooks, so drop the remainder. She can buy* 8.

6 $5{,}392 \approx 5{,}400$ *and* $88 \approx 90$. *So* $5{,}400 \div 90 = 60$.

7 $20.00 - 12.49 = 7.51$.

8 $108 \times 5 = 540$. *Two decimal places:* $540 \to 5.40$.

9 $a \div 0.1 = a \times 10$. *For example,* $3 \div 0.1 = 30$ *and* $3 \times 10 = 30$.

Find more at
ViewMath.com/ND-Grade5

10. LCD of 5 and 4 is 20. $\frac{3}{5} = \frac{12}{20}$ and $\frac{1}{4} = \frac{5}{20}$. So $\frac{12}{20} + \frac{5}{20} = \frac{17}{20}$.

11. $\frac{1}{9} \approx 0$, so $6\frac{1}{9} \approx 6$. $\frac{4}{5} \approx 1$, so $2\frac{4}{5} \approx 3$. Estimate: $6 - 3 = 3$.

12. LCD is 4. $\frac{1}{2} = \frac{2}{4}$. Borrow: $5\frac{2}{4} = 4\frac{6}{4}$. $\frac{6}{4} - \frac{3}{4} = \frac{3}{4}$. Wholes: $4 - 2 = 2$. Answer: $2\frac{3}{4}$.

13. A fraction $\frac{a}{b}$ means $a \div b$. So $\frac{3}{4} = 3 \div 4$.

14. $\frac{3}{7} \times 35 = \frac{105}{7} = 15$ students.

15. Amy added $4 + 3 = 7$ instead of multiplying $4 \times 3 = 12$. The correct answer is $\frac{1}{12}$.

16. $\frac{1}{4} < \frac{3}{4} < \frac{5}{4}$, so the products increase in the same order.

17. $\frac{10}{3} \times \frac{9}{4} = \frac{90}{12} = \frac{15}{2} = 7\frac{1}{2}$.

18. $\frac{2}{5} \times 30 = \frac{60}{5} = 12$ students.

19. $\frac{1}{2} \div 4 = \frac{1}{8}$.

20. Always start with the innermost grouping symbol. The parentheses $(10 - 6)$ are innermost, so subtract first.

21. An expression has no equals sign. Only B, $(3 \times 2) + 1$, has no equals sign.

22. The 5th terms are A: 4 and B: 16. The ordered pair is $(4, 16)$.

23. The x-axis is the horizontal number line that goes left to right.

Find more at
ViewMath.com/ND-Grade5

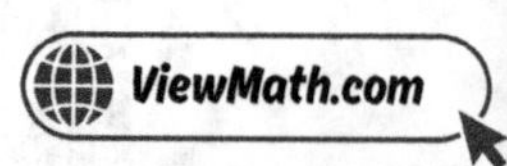

24 At 3 hours the cyclist should have ridden $3 \times 4 = 12$ miles, not 10.

25 $100 \times 3 = 300$ ft.

26 $0 + \frac{1}{4} + \frac{1}{4} + \frac{1}{2} + \frac{3}{4} + 0 + \frac{1}{2} = \frac{0+1+1+2+3+0+2}{4} = \frac{9}{4} = 2\frac{1}{4}$.

27 $4 \times 4 \times 4 = 64$ cubic units.

28 Bottom: $8 \times 6 \times 4 = 192$. Top: $4 \times 6 \times 3 = 72$. Total: $192 + 72 = 264$

29 $V = 18 \times 12 \times 8 = 1{,}728$ in^3.

30 Some parallelograms (like rectangles) have right angles, but many do not. The other statements are always true for their shape category.

✅ Practice Test 8 — Answer Key

1	C	2	C	3	3.00	4	B	5	B	6	B	7	A	8	1.50	9	C	10	B
11	C	12	B	13	B	14	C	15	B	16	Greater than 9	17	B	18	D	19	C
20	D	21	B	22	C	23	C	24	B	25	10,000 lb	26	B	27	2 layers	28	B
29	C	30	B																

Find more at
ViewMath.com/ND-Grade5

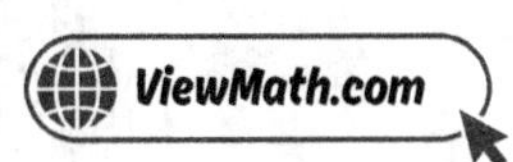

💡 Time to Learn! 💡

*Go through the explanations below, **especially for the questions you missed**.*

Understanding why each answer is correct makes you a stronger math thinker!

👍 *Tip: Circle any questions you got wrong, then read their explanation carefully.*

📖 Practice Test 8 — Detailed Explanations

1. $0.006 \times 10{,}000 = 60$, and $10{,}000 = 10^4$. The decimal point moves 4 places to the right: $0.006 \to 60$.

2. "Two hundred" is the whole-number part (200). The word "and" marks the decimal point. "Five hundredths" is 0.05. Together: 200.05.

3. The hundredths digit is 9. The thousandths digit is $9 \geq 5$, so round up: $9 + 1 = 10$ hundredths. Carry: $2.99 + 0.01 = 3.00$.

4. The tens digit is 3, which represents 30. $48 \times 30 = 1{,}440$.

5. $530 \div 24 = 22$ R2. The remainder (2) is the number of toys in the partially filled crate.

6. $5{,}520 \approx 5{,}600$ and $68 \approx 70$. So $5{,}600 \div 70 = 80$.

7. $50.00 - 32.99 = 17.01$.

8. $25 \times 6 = 150$. Two decimal places: $150 \to 1.50$.

9. Dividing by a number less than 1 gives a quotient larger than the dividend.

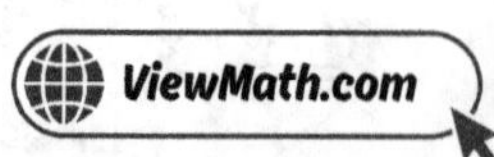

10. Multiples of 4: $4, 8, 12, 16, 20, \ldots$ Multiples of 10: $10, 20, \ldots$ The LCD is 20.

11. $\frac{5}{6} \approx 1$ and $\frac{2}{9} \approx 0$. Estimate: $1 - 0 = 1$.

12. "In all" means add. Wholes: $3 + 2 = 5$. LCD is 4. $\frac{1}{2} = \frac{2}{4}$. $\frac{2}{4} + \frac{3}{4} = \frac{5}{4} = 1\frac{1}{4}$. Regroup: $5 + 1\frac{1}{4} = 6\frac{1}{4}$.

13. $1 \div 3 = \frac{1}{3}$. Each child gets $\frac{1}{3}$ of the sandwich.

14. $\frac{5 \times 4}{8} = \frac{20}{8} = \frac{5}{2} = 2\frac{1}{2}$.

15. $\frac{2 \times 4}{3 \times 5} = \frac{8}{15}$.

16. $\frac{7}{3} > 1$ (since $7 > 3$), so the product is greater than 9.

17. Convert: $2\frac{2}{5} = \frac{12}{5}$. Multiply: $\frac{3}{4} \times \frac{12}{5} = \frac{36}{20} = \frac{9}{5} = 1\frac{4}{5}$.

18. "$\frac{1}{4}$ of 12" means multiply: $\frac{1}{4} \times 12 = 3$ slices.

19. $\frac{1}{5} \div 2 = \frac{1}{5 \times 2} = \frac{1}{10}$.

20. A: 20. B: $4 \times 5 = 20$. C: $10 \times 2 = 20$. D: $2 + 15 = 17$, which is not 20.

21. $325 + 18$ is larger than 325, so multiplying it by 7 gives a larger result.

22. $6 \times 5 = 30$.

23. In the first quadrant, the y-coordinate tells how far to move up from the x-axis.

Find more at
ViewMath.com/ND-Grade5

24 A steeper line shows that the y-values increase faster for each step in x.

25 $3\ T = 6{,}000$ *lb.* $6{,}000 + 4{,}000 = 10{,}000$ *lb.*

26 Heaviest: $\frac{1}{2}$. Lightest: $\frac{1}{4}$. Difference: $\frac{1}{2} - \frac{1}{4} = \frac{1}{4}$ *lb.*

27 Bottom: $6 \times 4 = 24$. Height: $48 \div 24 = 2$ layers.

28 Volume is additive because the total volume of a combined shape equals the sum of its non-overlapping parts.

29 $V = 25 \times 10 \times 2 = 500\ m^3$.

30 A rectangle with all 4 sides equal IS a square. But most rectangles have sides of different lengths and are NOT squares.

✅ Practice Test 9 — Answer Key

Find more at
ViewMath.com/ND-Grade5

💡 **Time to Learn!** 💡

Go through the explanations below, **especially for the questions you missed**.

Understanding why each answer is correct makes you a stronger math thinker!

👍 **Tip:** Circle any questions you got wrong, then read their explanation carefully.

📖 Practice Test 9 — Detailed Explanations

1. $10^2 = 100$. So $23 \times 100 = 2{,}300$. *The decimal point moves 2 places to the right.*

2. $6 \times 100 = 600$, $9 \times 1 = 9$, *total whole part* $= 609$. *Then* $5 \times \frac{1}{100} = 0.05$ *and* $3 \times \frac{1}{1{,}000} = 0.003$, *so decimal part* $= 0.053$. *Standard form:* 609.053.

3. *The hundredths digit is 7. Look at the thousandths digit (5). Since* $5 \geq 5$, *round up:* $7 + 1 = 8$. *Answer:* 46.88.

4. $3{,}050 \times 4 = 12{,}200$ *and* $3{,}050 \times 20 = 61{,}000$. $12{,}200 + 61{,}000 = 73{,}200$.

5. $100 \div 9 = 11$ R1. *There is 1 foot of rope left over.*

6. $47 \approx 50$ *and* $63 \approx 60$. *So* $50 \times 60 = 3{,}000$.

7. *Write 6 as* 6.00. *Then* $6.00 - 2.45 = 3.55$.

8. *Ignore the decimal:* $615 \times 2 = 1{,}230$. *Two decimal places, so* $1230 \rightarrow 12.30$.

9. *Multiply both by 10:* $81.2 \div 4 = 20.3$.

Find more at
ViewMath.com/ND-Grade5

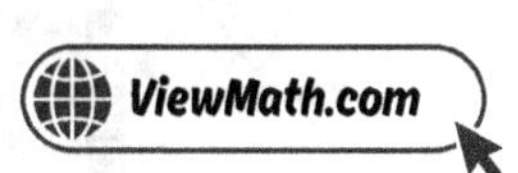

10   LCD is 9. $\frac{2}{3} = \frac{6}{9}$. So $\frac{6}{9} + \frac{5}{9} = \frac{11}{9} = 1\frac{2}{9}$.

11 $\frac{7}{8} \approx 1$ and $\frac{4}{5} \approx 1$. Estimate: $1 + 1 = 2$. Since $\frac{11}{13} < 1$, the answer is far too small and not reasonable.

12   Add. Wholes: $1 + 2 = 3$. LCD of 2 and 3 is 6. $\frac{1}{2} = \frac{3}{6}$ and $\frac{1}{3} = \frac{2}{6}$. $\frac{3}{6} + \frac{2}{6} = \frac{5}{6}$. Total: $3\frac{5}{6}$.

13 $11 \div 4 = 2$ remainder 3, so $\frac{11}{4} = 2\frac{3}{4}$. It lies between 2 and 3.

14 $\frac{5 \times 3}{6} = \frac{15}{6} = 2\frac{3}{6} = 2\frac{1}{2}$.

15 $\frac{2}{3} \times \frac{5}{6} = \frac{10}{18} = \frac{5}{9}$ square foot.

16   She multiplies by $\frac{2}{3}$, which is less than 1, so she uses less flour.

17 $\frac{8}{3} \times \frac{9}{8} = \frac{72}{24} = 3$.

18 $\frac{5}{6} \times 6 = \frac{30}{6} = 5$ miles.

19 $\frac{1}{4} \div 3 = \frac{1}{12}$, so the missing number is 3.

20   Innermost first: $2 + 1 = 3$. Brackets: $8 - 3 = 5$. Then $3 \times 5 = 15$.

21   Cost of notebooks: $3 \times 4 = 12$. Money left: $20 - 12 = 8$. Expression: $20 - (3 \times 4)$.

22   A: $0, 2, 4, 6$. B: $0, 10, 20, 30$. The 4th pair is $(6, 30)$.

23   The x-coordinate (4) tells how far right. The y-coordinate (7) tells how far up.

Find more at
ViewMath.com/ND-Grade5

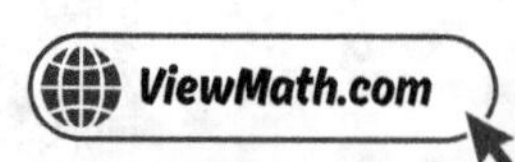

24 $7 \times 2 = 14$ cm.

25 1 qt = 2 pt, so $6 \times 2 = 12$ pt.

26 When using eighths, the tick marks should show every eighth: $\frac{1}{8}, \frac{2}{8}, \frac{3}{8}, \frac{4}{8}$ (or equivalently $\frac{1}{4}, \frac{3}{8}, \frac{1}{2}$).

27 $40 \div 8 = 5$ layers.

28 Full: $10 \times 10 \times 6 = 600$. Removed: $5 \times 5 \times 6 = 150$. Remaining: $600 - 150 = 450$ cm^3.

29 $V = l \times w \times h$. $96 = 8 \times w \times 2 = 16w$. $w = 96 \div 16 = 6$ ft.

30 A square has 4 right angles (rectangle property) AND 4 equal sides (rhombus property).

☑ Practice Test 10 — Answer Key

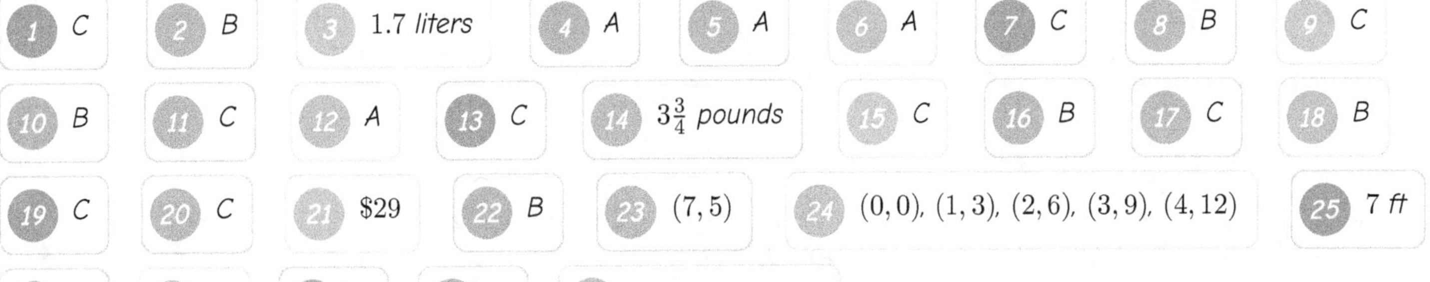

1 C	2 B	3 1.7 liters	4 A	5 A	6 A	7 C	8 B	9 C
10 B	11 C	12 A	13 C	14 $3\frac{3}{4}$ pounds	15 C	16 B	17 C	18 B
19 C	20 C	21 $29	22 B	23 (7,5)	24 (0,0), (1,3), (2,6), (3,9), (4,12)	25 7 ft
26 C	27 C	28 A	29 B	30 Sometimes true

Find more at
ViewMath.com/ND-Grade5

💡 **Time to Learn!** 💡

Go through the explanations below, **especially for the questions you missed**.

Understanding why each answer is correct makes you a stronger math thinker!

👍 *Tip:* Circle any questions you got wrong, then read their explanation carefully.

📖 *Practice Test 10 — Detailed Explanations*

1 If $? \div 100 = 3.7$, then $? = 3.7 \times 100 = 370$. Dividing moves the decimal left, so undoing the division means multiplying.

2 "Eight hundred" is the whole part (800). "And" is the decimal point. "Twelve thousandths" is 0.012. Together: 800.012.

3 The tenths digit is 6. The hundredths digit is $5 \geq 5$, so round up: $6 + 1 = 7$. Answer: 1.7 liters.

4 $415 \times 8 = 3{,}320$ and $415 \times 30 = 12{,}450$. $3{,}320 + 12{,}450 = 15{,}770$.

5 $350 \div 24 = 14 \text{ R}14$. The remainder is 14, so there are 14 cookies in the partially filled bag.

6 $88 \approx 90$ and $115 \approx 100$. So $90 \times 100 = 9{,}000$.

7 Write 4 as 4.0 (or 4.00) so the decimal points line up. $4.0 - 1.7 = 2.3$.

8 $4.7 \approx 5$. $5 \times 8 = 40$. The exact answer is 37.6, which is close to 40.

9 Multiply both by 100: $500 \div 25 = 20$. There are 20 quarters in \$5.00.

Find more at
ViewMath.com/ND-Grade5

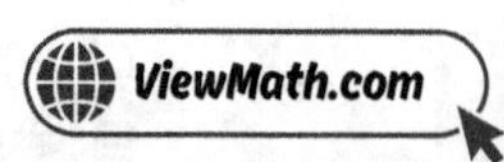

10. LCD of 3 and 4 is 12. $\frac{2}{3} = \frac{8}{12}$ and $\frac{1}{4} = \frac{3}{12}$. So $\frac{8}{12} + \frac{3}{12} = \frac{11}{12}$.

11. The numerator 7 is very close to the denominator 8, so $\frac{7}{8}$ is close to 1.

12. Total eaten: LCD of 3, 4, and 6 is 12. $\frac{1}{3} = \frac{4}{12}$, $\frac{1}{4} = \frac{3}{12}$, $\frac{1}{6} = \frac{2}{12}$. Sum $= \frac{9}{12} = \frac{3}{4}$. Left: $1 - \frac{3}{4} = \frac{1}{4}$.

13. 3 pizzas $\div$ 4 friends $= \frac{3}{4}$ of a pizza each.

14. $\frac{3}{4} \times 5 = \frac{15}{4} = 3\frac{3}{4}$ pounds.

15. $\frac{2 \times 3}{3 \times 2} = \frac{6}{6} = 1$.

16. $\frac{6}{5} > 1$ (since $6 > 5$), so he will run more than 5 miles.

17. Convert: $3\frac{1}{2} = \frac{7}{2}$. Then $\frac{4}{1} \times \frac{7}{2} = \frac{28}{2} = 14$.

18. $\frac{5}{8} \times 48 = \frac{240}{8} = 30$ cars.

19. $\frac{1}{3} \div 4 = \frac{1}{3 \times 4} = \frac{1}{12}$.

20. Parentheses first: $2 + 6 = 8$. Then $48 \div 8 = 6$.

21. $(8 \times 3) + 5 = 24 + 5 = 29$.

22. $24 \div 8 = 3$. B is 3 times A.

23. 7 right = x-coordinate 7, 5 up = y-coordinate 5. His location is $(7, 5)$.

Find more at
ViewMath.com/ND-Grade5

24   3 *cups per day:* $0, 3, 6, 9, 12$. *Pair with days* $0, 1, 2, 3, 4$

25 $84 \div 12 = 7$ *ft.*

26 $3 + 5 + 2 + 1 = 11$ *worms.*

27   *A unit cube with 1 cm edges has a volume of 1 cubic centimeter* (1 cm^3).

28   *Full:* $12 \times 6 \times 4 = 288$. *Notch:* $4 \times 2 \times 4 = 32$. *Remaining:* $288 - 32 = 256$ in^3.

29   *Estimate:* $10 \times 10 \times 4 = 400$. *Exact:* $11 \times 9 \times 4 = 396$, *which is close to* 400.

30   *Some parallelograms (those with 4 right angles) are rectangles, but slanted parallelograms are not.*

Great job checking your work!

Keep practicing and you'll be a math star!

Find more at
ViewMath.com/ND-Grade5

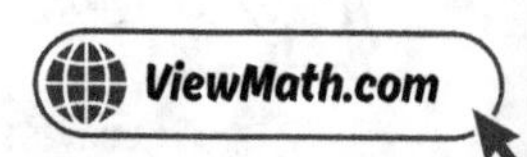

Author's Final Note

I hope you enjoyed this book as much as I enjoyed writing it. Whether you are a student working through the material, a parent supporting your child's learning, or a teacher guiding your class, I have tried to make this book as clear and engaging as possible. I hope I have succeeded. If you have any suggestions for improvement, please let me know. I would love to hear from you.

The accuracy of calculations is very important to me. We have done our best, but I also expect that I have made some minor errors. Constant improvement is the name of the game. If you find any errors, please let me know. I will fix them in the next edition.

For students: Your learning journey does not end here. I have written a series of books to help you learn math. Make sure you browse through them. I especially recommend workbooks and practice tests to help you prepare for your exams.

For parents: Thank you for investing in your child's education. I encourage you to explore the companion resources available online to help support your child outside the classroom.

For teachers: Thank you for the invaluable work you do every day. I hope this book serves as a useful resource in your classroom. Feel free to reach out if you have suggestions or would like to discuss how best to use this book with your students.

I also enjoy reading your reviews. If you have a moment, please leave a review on where you found this book. It will help others find this book. If you have any questions or comments, please feel free to contact me at drNazari@ViewMath.com.

And one last thing: Remember to use online resources for additional help. I recommend using the resources on `https://ViewMath.com` You can find video lessons, practice problems, and more. You can also use the online companion for this book to track your progress and access additional resources.

Wishing all students the best in their studies, parents every success in supporting their children, and teachers continued inspiration in their classrooms!

Dr. A. Nazari

 ## Great Job! Keep Learning with ViewMath!

Keep up the great work! Visit **viewmath.com/ND-Grade5** for free lessons, quizzes, and more.

Study Guide

Workbook

Step-by-Step

3 Practice Tests

5 Practice Tests

ViewMath.com